Advances in Nanostructured Electrode Materials: Design and Applications

Advances in Nanostructured Electrode Materials: Design and Applications

Guest Editor

Maria Grazia Musolino

MDPI

Basel • Beijing • Wuhan • Barcelona • Belgrade • Novi Sad • Cluj • Manchester

Guest Editor
Maria Grazia Musolino
Department of Civil, Energy,
Environmental and Materials
Engineering (DICEAM)
Mediterranean University
of Reggio Calabria
Reggio Calabria
Italy

Editorial Office
MDPI AG
Grosspeteranlage 5
4052 Basel, Switzerland

This is a reprint of the Special Issue, published open access by the journal *Nanomaterials* (ISSN 2079-4991), freely accessible at: https://www.mdpi.com/journal/nanomaterials/special_issues/EER41U2O7R.

For citation purposes, cite each article independently as indicated on the article page online and as indicated below:

Lastname, A.A.; Lastname, B.B. Article Title. *Journal Name* **Year**, *Volume Number*, Page Range.

ISBN 978-3-7258-6392-1 (Hbk)
ISBN 978-3-7258-6393-8 (PDF)
https://doi.org/10.3390/books978-3-7258-6393-8

Contents

About the Editor

Maria Grazia Musolino

Maria Grazia Musolino is currently Associate Professor of Chemical Fundamentals of Technologies at the Mediterranean University of Reggio Calabria since 2007. She earned her master's degree, cum laude, in Chemistry from the University of Messina in 1991 and her PhD degree in "Chemistry of Materials for Special Uses" in 1997 along with a Specialization degree in "Chemical Process Technologies" in the same year. After two years of post-doctoral research (1999–2001) spent at the Mediterranean University of Reggio Calabria, she was appointed Assistant Professor at the same university in 2001, during which she also spent a period as a "visiting researcher" at the University of British Columbia in Vancouver. Her research focuses mainly on sustainability issues, in particular the synthesis and characterization of (nano)materials for catalytic and energy-related applications. Her scientific interests include: the development and characterization of graphene-based nanostructured composites and high entropy materials for applications as electrode materials; the development and characterization of supported mono- and bi-metallic catalysts and their use in important reactions of green chemistry; the study of the reaction mechanisms in heterogeneous catalysis. She has authored over fifty publications in peer-reviewed international scientific journals and has contributed to national and international research projects.

nanomaterials

MDPI

Editorial

Advances in Nanostructured Electrode Materials: Design and Applications

Maria Grazia Musolino [1,2]

[1] Department of Civil, Energy, Environmental and Materials Engineering (DICEAM), Mediterranean University of Reggio Calabria, 89122 Reggio Calabria, Italy; mariagrazia.musolino@unirc.it

[2] National Reference Center for Electrochemical Energy Storage (GISEL), National Interuniversity Consortium for the Science and Technology of Materials (INSTM), 50121 Florence, Italy

The development of nanostructured electrode materials is a cornerstone of emerging electrochemical technologies that provides clean and sustainable solutions to address the global energy demand and the rapid depletion of fossil fuels, as well as environmental pollution issues such as global warming, climate change, and air/water pollution resulting from the overexploitation of carbon-based energy resources. The unique and outstanding features of these materials, arising from the ability to suitably tailor their structural and functional properties, are key to optimizing performance, durability, and efficiency. However, the design, synthesis, and characterization of nanostructured electrodes remain a current area of research, facing challenges such as scalable fabrication and structural stability.

The Special Issue of *Nanomaterials* "Advances in Nanostructured Electrode Materials: Design and Applications" provides an overview of the current state of the art in the field of nanostructured electrode materials, including their synthesis, properties, characterization, and applications in electrocatalysis, energy conversion, energy storage, and environmental protection. It especially promotes the interdisciplinary aspects of materials science and the interactions between fundamental research and technology. This Special Issue collects eight original research papers, summarized below, covering a wide range of electrode materials and their technological applications.

Cai et al. investigated the effects of introducing elemental tungsten (W) into GeTe-based materials on the material structure and electrical and thermal transport properties [1]. Experimental data indicated that the increase in electrical conductivity of the sample was due to the presence of high-valence state W atoms, additional charge carriers, thus improving the Seebeck coefficient. Furthermore, appropriate W doping concentrations reduced the lattice thermal conductivity by optimizing the material structure, thus comprehensively tuning the thermoelectric transport properties of GeTe systems.

Conti et al. synthesized a self-standing electrode for sodium-ion batteries (SiBs), based on a $Na_3MnTi(PO_4)_3$ active material loaded into carbon nanofibers (CNFs), by electrospinning [2]. The active material was homogeneously spread into CNFs and displayed a NASICON-type crystal structure, but the high sintering temperature (750 °C) used to obtain conductive CNFs induced cell shrinkage, thus implying a sluggish redox activity. This study highlighted the promising electrochemical performances of this electrode compared to its conventional tape-casted counterpart, thanks to the easy electrolyte diffusion and contact with the active material afforded by the porous nature of non-woven nanofibers.

The work of Beitia et al. focused on the synthesis, characterization, and electrochemical performance of zinc-doped manganese hexacyanoferrate systems, used as cathodes for aqueous Zn-ion batteries [3]. Zinc doping was a useful strategy to improve the weak structural stability of manganese hexacyanoferrate (MnHCF) in an aqueous environment

 https://doi.org/10.3390/nano15211638

and reduce manganese dissolution. Despite a decrease in the specific capacity of the system, $Mn_{1-x}Zn_xHCF$ (x = 0, 0.25, 0.5, 0.75 and 1) Prussian Blue analogues, prepared through a simple and easy-to-implement approach, improved the stability and capacity retention of the cathode. Furthermore, the amount of zinc introduced into the MnHCF played a critical role in achieving higher reversibility and stable performance for the MnHCF phase without a drastic loss in capacity.

A novel design concept, where paper can act as both a separator and a substrate for coating the anode material for lithium-ion batteries, is reported in the work of Blomquist et al. [4]. A fully disposable and resource-efficient paper-based electrode was successfully fabricated via large-scale roll-to-roll coating technology, where the conductive material is a nanographite and microcrystalline cellulose mixture coated on a paper separator. The produced electrode material, tested in a typical lithium ion half-cell coin cell setup, exhibited a specific capacity of 147 mAh/g and a good long-term stability of the battery capacity over extended cycling.

Gasparotto et al. developed the chemical vapor infiltration (CVI) method for the one-step synthesis of polymeric carbon nitride (PCN) films on porous Ni foam substrates, starting from melamine as a precursor compound [5]. By varying the reaction temperature and the precursor amount, PCN deposits with a tunable condensation degree (from melem/melon hybrids to melon-like materials) and different morphological features are obtained. PCN-based electrodes with different polymerization degrees showed promising catalytic performances in the oxygen evolution reaction (OER), considered a bottleneck of the water splitting process.

A new type of Metal–Organic Framework (MOF) material (metal-triazolates) on nickel foam (NF) substrates, synthesized by a solvothermal method, was directly used as a self-supporting electrode for the OER, enabling performances that outperform most of the reported OER catalysts [6]. Among various metal-triazolates, the Fe-based one on (MET-Fe/NF) exhibited the best OER performance, achieving a low overpotential of 122 mV at a current density of 10 $mAcm^{-2}$ and maintaining good stability over 15 h. The experimental results showed that MET-Fe/NF underwent structural reconstruction during the OER process, resulting in a hybrid catalyst with several active components (iron/nickel (oxy)hydroxides) with high OER activity. Furthermore, in a two-electrode water splitting setup, this electrocatalyst, used as an anode, displayed a good performance, allowing for continuous hydrogen and oxygen generation at a low voltage of 1.46 V.

Sako et al. fabricated Ni nanowire array electrodes with an extremely large surface area by a potentiostatic electrodeposition technique into anodized alumina nanochannels [7]. The electrodeposited Ni nanowire arrays had a textured structure with a preferential orientation in the fcc-Ni (111) plane, regardless of the electrodeposition potential and exhibited uniaxial magnetic anisotropy, with easy magnetization in the axial direction. The electrocatalysts exhibited a lower overpotential and a higher current density towards the hydrogen evolution reaction compared to electrodeposited Ni films.

Campagna Zignani et al. proposed a very simple and scalable process for the synthesis of low cost and efficient bimetallic oxide-based electrocatalysts for "green hydrogen" production from water electrolysis [8]. Nanostructured NiCo- and NiFe-based electrode materials with different Ni molar fractions were prepared by the sol–gel method and subsequent calcination in air at different temperatures, and then evaluated as anode materials in a zero-gap anion exchange membrane water electrolysis (AEMWE) full cell. The cell cathodes were fabricated using the same materials after reduction in a H_2/Ar atmosphere. The electrochemical results revealed that the nanomaterial phase purity and the average crystal size were critical in determining cell performance. Highly pure and finely grained electrocatalysts yielded higher current densities at lower overpotentials, paving the way for scalable and cost-effective green hydrogen production from water electrolysis.

In conclusion, the research presented in this Special Issue underscores the immense potential of nanostructured electrode materials in wide range of applications. Each paper provides valuable insights into the synthesis, characterization, and application of these materials, showcasing their versatility. Although this Special Issue cannot fully cover the topic of electrode materials, I firmly believe that its contributions will open new perspectives in the field of materials and technologies, promoting the development of advanced nanomaterials as electrodes, combining low-cost, sustainable materials with simple manufacturing processes, in line with future clean energy goals.

Funding: This research received no external funding.

Acknowledgments: I am grateful to all the authors for submitting their work to the Special Issue and for its successful completion. Special thanks to all the reviewers for their constructive suggestions aimed at improving the quality and impact of the submitted manuscripts. Finally, I sincerely thank Colleen Zhang and all the editorial staff of *Nanomaterials* for their support in making the entire creative process of this Special Issue smooth and efficient.

Conflicts of Interest: The author declares no conflicts of interest.

References

1. Cai, Z.; Zheng, K.; Ma, C.; Fang, Y.; Ma, Y.; Deng, Q.; Li, H. Ultra-Low Thermal Conductivity and Improved Thermoelectric Performance in Tungsten-Doped GeTe. *Nanomaterials* **2024**, *14*, 722. [CrossRef] [PubMed]
2. Conti, D.M.; Urru, C.; Bruni, G.; Galinetto, P.; Albini, B.; Berbenni, V.; Girella, A.; Capsoni, D. $Na_3MnTi(PO_4)_3$/C Nanofiber Free-Standing Electrode for Long-Cycling-Life Sodium-Ion Batteries. *Nanomaterials* **2024**, *14*, 804. [CrossRef] [PubMed]
3. Beitia, J.; Ahedo, I.; Paredes, J.I.; Goikolea, E.; Ruiz De Larramendi, I. Exploring Zinc-Doped Manganese Hexacyanoferrate as Cathode for Aqueous Zinc-Ion Batteries. *Nanomaterials* **2024**, *14*, 1092. [CrossRef] [PubMed]
4. Blomquist, N.; Phadatare, M.; Patil, R.; Zhang, R.; Leuschen, N.; Hummelgård, M. Large-Scale Compatible Roll-to-Roll Coating of Paper Electrodes and Their Compatibility as Lithium-Ion Battery Anodes. *Nanomaterials* **2025**, *15*, 113. [CrossRef] [PubMed]
5. Gasparotto, A.; Barreca, D.; Maccato, C.; Pierobon, E.; Rizzi, G.A. One-Step Synthesis of Polymeric Carbon Nitride Films for Photoelectrochemical Applications. *Nanomaterials* **2025**, *15*, 960. [CrossRef]
6. Cai, K.; Chen, W.; Wan, Y.; Chu, H.; Hai, X.; Zou, R. Self-Reconstructed Metal–Organic Framework-Based Hybrid Electrocatalysts for Efficient Oxygen Evolution. *Nanomaterials* **2024**, *14*, 1168. [CrossRef] [PubMed]
7. Sako, Y.; Saeki, R.; Hayashida, M.; Ohgai, T. Uniaxial Magnetization and Electrocatalytic Performance for Hydrogen Evolution on Electrodeposited Ni Nanowire Array Electrodes with Ultra-High Aspect Ratio. *Nanomaterials* **2024**, *14*, 755. [CrossRef]
8. Campagna Zignani, S.; Fazio, M.; Pascale, M.; Alessandrello, C.; Triolo, C.; Musolino, M.G.; Santangelo, S. Cost-Effective Bimetallic Catalysts for Green H_2 Production in Anion Exchange Membrane Water Electrolyzers. *Nanomaterials* **2025**, *15*, 1042. [CrossRef] [PubMed]

nanomaterials

MDPI

Article

Large-Scale Compatible Roll-to-Roll Coating of Paper Electrodes and Their Compatibility as Lithium-Ion Battery Anodes

Nicklas lomquist *, Manisha Phadatare, Rohan Patil, Renyun Zhang, Noah Leuschen and Magnus Hummelgård

Department of Engineering, Mathematics and Science Education, Mid Sweden University, SE-851 70 Sundsvall, Sweden; manisha.phadatare@miun.se (M.P.); rohan.patil@miun.se (R.P.); renyun.zhang@miun.se (R.Z.); magnus.hummelgard@miun.se (M.H.)

* Correspondence: nicklas.blomquist@miun.se

Abstract: A recyclability perspective is essential in the sustainable development of energy storage devices, such as lithium-ion batteries (LIBs), but the development of LIBs prioritizes battery capacity and energy density over recyclability, and hence, the recycling methods are complex and the recycling rate is low compared to other technologies. To improve this situation, the underlying battery design must be changed and the material choices need to be made with a sustainable mindset. A suitable and effective approach is to utilize bio-materials, such as paper and electrode composites made from graphite and cellulose, and adopt already existing recycling methods connected to the paper industry. To address this, we have developed a concept for fabricating fully disposable and resource-efficient paper-based electrodes with a large-scale roll-to-roll coating operation in which the conductive material is a nanographite and microcrystalline cellulose mixture coated on a paper separator. The overall best result was achieved with coated roll 08 with a coat weight of 12.83(22) g/m^2 and after calendering, the highest density of 1.117(97) g/cm^3, as well as the highest electrical conductivity with a resistivity of 0.1293(17) $m\Omega \cdot m$. We also verified the use of this concept as an anode in LIB half-cell coin cells, showing a specific capacity of 147 mAh/g, i.e., 40% of graphite's theoretical performance, and a good long-term stability of battery capacity over extended cycling. This concept highlights the potential of using paper as a separator and strengthens the outlook of a new design concept wherein paper can both act as a separator and a substrate for coating the anode material.

Keywords: paper electrodes; sustainable; recyclable; resource efficient; graphene; nanographite; nanoplatelets; cellulose binder; energy storage; lithium ion

1. Introduction

The rapid growth of electric vehicles (EVs) and growing concern over global warming, reducing the carbon emissions associated with the entire life cycle of lithium-ion batteries (LIBs) have become a central focus for various entities including research institutions across the globe [1]. Developing LIBs with a recyclability perspective is essential for reducing carbon dioxide emissions. However, today's lithium-ion batteries prioritize battery capacity and energy density over recyclability, and hence, lithium batteries are currently being recycled at rates less than 50%, as compared with lead–acid batteries that have a recycling rate of 99.5% [2]. When it comes to recycling, the emphasis should be on developing recycling methods that allow the entire battery to be recycled. However, most of efforts are being made to improve recycling methods associated with cathode materials, while other critical components such as electrolytes that are often toxic to the environment do not receive any major attention, unfortunately [3]. Step utilization and recovery separation reproduction are the two approaches used to increase sustainability; in short, the former makes more efficient use of existing batteries, and the latter is a cluster of different types of recycling methods, i.e., pyro-, hydro-, and biometallurgy, and direct regeneration [3]. Neither of these methods is ideal and all have some major drawbacks; pyrometallurgical processes are simpler but suffer from not being able to recover the lithium metal and releasing hazardous gases [4]. Hydrometallurgical treatment, which is the most popular one today with more than half of all processes sorted under this category, also allows for lithium recovery but is more complex than the pyrometallurgical process [4]. To achieve good efficiency in recovering the metals, it has been proposed that hydrometallurgy should be combined with biometallurgy, followed by a second step of bioleaching with bacteria or funguses to recover the metals and boost the efficiency of the recovery process [3]. The drawback of the bioleaching process is that as the concentrations of minerals in the waste material go up, the bacteria's efficiency decreases due to the toxic effects [4]. The direct regeneration method implies that materials are not first recycled but, in some way, directly reused, i.e., the cells are repacked with old electrode materials, and although it is a simple and resource-effective approach, the drawbacks are contamination and performance issues of the regenerated cells [5]. The issues being mentioned are caused by design flaws in batteries where the focus has been on developing cells with high performance and high economical return. However, from a sustainable perspective, the development lacks the two remaining factors of high environmental benefit and high safety [6]. A sustainable-perspective overview of the components of a regular lithium-ion battery presents several issues. The current collectors, made from aluminum and copper foils, are produced via high-energy-consuming processes, and the cathode materials are high in cost, non-renewable, and contain toxic elements. The binders used are fluorinated versions, and electrolytes such as LiPF6 are toxic and unsafe in decomposition scenarios. The anode materials are graphite versions from non-renewable sources that contribute to either pollution or high energy consumption in their production. Lastly, the separator is made from plastic, is not thermally stable, i.e., a safety issue, and is produced via mineral oils [7].

As the quote says, if the shoe fits, so it is the underlying battery design that must be changed to more efficiently improve the situation. A new shoe, or rather a sustainable battery design, is to a high degree dominated by material choices. These materials must be able to fulfill these high environmental benefit and high safety criteria and do so at their core level and all the way throughout the value chain. A new and more effective approach is to utilize bio-materials in battery design more frequently, and there are many publications on electrodes where the fossil or synthetic graphite is replaced by biocarbon products. These are much more resource-efficient and can be generated from biomass or biowaste materials; lignin and cellulose are two sources for biocarbon generation [8]. These biomass materials also contain other biomolecules that in the long run can function as the active components in the redox reactions of the battery instead of relying on metals, making a completely metal-free battery [9]. The metallic current collector itself can also be replaced with carbon alternatives, eliminating the energy consumption of metal foil production; one way of achieving this is by the direct carbonization of a cellulose paper substrate in a furnace, or one can use laser processing tools to carbonize the top layer of a paper directly [10,11]. Using biobased recycling methods such as fungi or bacteria for recycling is an overall good and sustainable approach, as long as the environment the organisms operate in is suitable; this leads to less toxicity as well as good organism penetration and accessibility into the battery during decomposition due to the reduced usage of metals, foils, and plastics, which is needed to improve their overall slow performance [12]. This concept can be extended further by allowing the organisms to construct the new electrode material as well; this is referred to as biofabrication [13]. Hence, many of the components in a battery can be built around different types of paper substrates, and this gives us the concept of paper-based batteries, or paper-based energy storage devices in general, and these have potential to provide many solutions to the sustainability problems of existing battery technology. As described above, a cellulose-based material is renewable, resource-efficient, and non-toxic, has a fully developed recycling platform in its value chain, and lastly, it also has an existing industry capable of producing the large volumes of materials needed for a green societal transition via battery technology implementation [14]. Paper can serve as a separator instead of plastic, and it can be converted into electrodes by different coating techniques. It can achieve current collector functionality through carbonization methods, and since the material is flexible, new types of packaging and encapsulating techniques can also be introduced, such as folding [15]. LIB polypropylene (PP) or polyethene (PE) separators are usually less than 25 μm thick sheets with 40% porosity and sub-micrometer pore sizes. Using paper as a separator has been shown to be advantageous even in terms of performance, as these have proven MacMullin numbers (ratio of ion conductivity between a wetted separator and the ion conductivity of the free electrolyte) of 3–6 [16], compared to the typical 20 for a PE separator [17], where a lower ratio corresponds to better ion conductivity. To make a paper-based electrode, the paper itself can be converted into an electrode, and this has already been demonstrated in pilot-scale operations in which paper was fabricated containing active carbon and PEEDOT:PSS ingredients [18,19]. However, most often, regular coating techniques or methods are used, such as printing, casting/filtration, or thermal evaporation to deposit an electrically conductive material on top of the paper to serve as the electrode. Commonly, different types of nanomaterial composites with carbon nanotubes, graphene, or similar are used, and in this scenario, paper obtains additional functionality to serve as a separator as well [20]. Adding conductive materials such as carbon nanotubes on top of the paper via Meyer rod coating has already demonstrated excellent results, such as achieving a surface resistivity of 1 ohm per square, and realized devices such as supercapacitors with specific capacitance of 200 F/g [21]. Another reported method is the spray deposition of graphite and microfibrillated cellulose onto bleached softwood pulp prior to pressing and drying in a pilot paper machine. This method achieves an LIB anode capacity of 95 mAh/g at 1 C with an electrode thickness of 27.5 μm, specific weight of 11.6 g/m^2, and electrical resistivity of about 500 Ω/sq (about 14 Ω·m) [22]. Besides paper electrodes being used for supercapacitor purposes, lithium-ion paper batteries have already been demonstrated with cellulose binders in a $LiFePO_4$ cathode and graphite anode [14,23]. Even more complex electronic devices can be realized with paper; for example, printing allows us to print a battery on a local area of the paper, which can be integrated with other surrounding components [24]. In addition to energy storage applications, our own studies have shown that paper with conductive coatings can also be used for energy harvesting in triboelectric nanogenerators with power densities exceeding 14 kWm^{-2} [25].

Here, we report the results of fabricating a fully disposable and resource-efficient paper-based electrode with a large-scale roll-to-roll coating operation in a paper pilot facility at speeds of up to 25 m/min in which the conductive material is a graphene/graphite mixture (nanographite) with microcrystalline cellulose (MCC) as the binder. The nanographite was fabricated in-house by our earlier-developed water-based large-scale compatible exfoliation technique [26,27]. The produced electrode material was then verified in a typical lithium-ion half-cell coin cell setup in which the electrode paper acted as both an anode and separator, together with standard lithium foil as the cathode, and default LP40 was used as the electrolyte.

2. Materials and Methods

2.1. Electrode Material Preparation

In this study, two different coating colors (electrode material suspensions) were used, called Slurry A and Slurry B. Both coating colors had the same amount of MCC addition as the binder but different nanographite sources. For Slurry A, a 1000 L nanographite suspension (GS14, SN:1013-1052) with 40 gL^{-1} solids content was purchased from 2Dfab in Sundsvall, Sweden. This nanographite was based on graphite raw material from Imerys and a Pluronic dispersion agent. The nanographite suspension was de-watered to 95 gL^{-1} solids content by removing the clear water phase above the sedimented nanographite. For Slurry B, a 400 L nanographite suspension was fabricated in-house by our own water-based large-scale compatible exfoliation technique, described earlier by Blomquist et al. [27]. The shear zone used was a 2 mm helical coil tube, referred to as S2 in the original method. The solids content of graphite in the suspension was increased to 75 gL^{-1} and the flow rate was held constant at 5 $Lmin^{-1}$ for 10 full passes through the shear zone. The graphite used was thermally expanded natural crystalline graphite (EXG 9840) from Graphit Kropfmühl in Passau, Germany with an addition of 2 wt% polyacrylic acid (Sigma-Aldrich, St. Louis, MO, USA), in relation to the graphite mass, as the dispersant. For additional details, see Supplementary Figure S1A–C. The MCC used as the binder was Refined MCC (White MCC) from Fibenol in Tallinn, Estonia. The MCC was delivered as a paste with 18.5% solids content and was used as is without further modifications. In both Slurry A and Slurry B, 10 wt% MCC was added to the nanographite suspensions, recalculated as dry weight in relation to the dry content of nanographite. The nanographite suspensions with MCC were mixed using a Cowles mixer to achieve uniform coating colors. The volume and solids contents of the final coating colors were 440 L at 10 wt% for Slurry A and 412 L at 8 wt% for Slurry B.

2.2. Electrode Coating

Two different substrate papers were used in this study: Advantage Kraft Plus 70 g (Plus), with a measured grammage of 68.54(22) g/m^2 and a thickness of 96.60(50) μm, and Advantage Boost HP 80 g (Boost), with a measured grammage of 78.87(19) g/m^2 and a thickness of 124.83(45) μm, both Kraft papers and both from Mondi Dynäs in Kramfors, Sweden. The substrates were selected based on their high dry and wet tensile strength and porosity. The main difference in properties, except the grammage, between the substrates is air permeability; for Advantage Boost HP, no hydrophobing agent is added in production, leading to higher permeability. The substrate papers were delivered as 500 kg rolls with a diameter of 1200 mm and a width of 520 mm. The rolls were used as is, and the coating was applied on the outer slightly rougher paper side of the rolls. The coating was performed at the UMV Coating Systems Pilot plant in Säffle, Sweden, using the UMV Liquid Application System (LAS). In short, this coating applicator transferred the coating color with an applicator roll from a pan to a metering nip. The metered film of coating color was then applied onto the substrate paper by a hydrophilic transfer roll. The variable coating parameters were machine speed, transfer roll speed, and metering nip, as well as individual control of a series of IR dryers and hot air dryers. See Supplementary Figure S2A–D for images of the coating operation.

2.3. Sample Preparation and Characterization

Samples from the coated rolls were cut both across the width of the roll and lengthwise according to Figure 1.

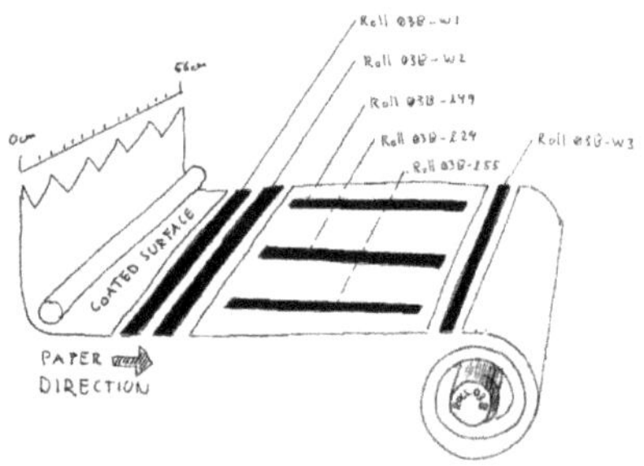

Figure 1. Schematic sketch of how sample strips were cut from rolls of coated paper, with examples of how the samples were named.

For each roll, the following samples were taken: Three sample strips with a width of 2–3 cm were cut across the full width of the roll; the two first strips were cut 10 cm apart from each other, and the third strip was cut 1 m away from the first strip; these are referred to as w1, w2, and w3. Another set of three sample strips, 1 m long and also about 2–3 cm wide, were cut in the lengthwise direction of the roll from three different sections of the roll width. The first of these sample strips was positioned 5.5 cm from the edge (edge-facing operator), the second was positioned 29 cm from the edge (the middle section), and the third sample strip was positioned 49 cm from the edge. These sample strips are referred to as l55, l29, and l49. Prefixes to notation correspond to the roll or sheet sample and roll number; for example, roll-03B-l55 refers to roll sample 03B, a lengthwise section 5.5 cm from the edge.

2.3.1. Electrical and Physical Measurements

For each sample strip, the electrical resistance was measured between the edge and multiple points along the length of the strip. The measurements were made in stepwise length increases of 1 cm with a Tillquist TQ711 multimeter (Kista, Sweden), and the resistance dependency as a function of length was determined with linear regression. The electrical bulk resistivity (ρ) of the coating was then calculated from the slope factor of the regression, as given by Equation (1):

$$R = \frac{\rho}{WH} \cdot L, \tag{1}$$

in which R is the measured electrical resistance, W is the width of the strip, H is the coating thickness, and L is the length of the strip. To measure the coating thickness and calculate coat weight, sets of 100 coins with a diameter of 16 mm were punched out from the sample strips including one strip of uncoated substrate paper as a reference. The thickness was measured with a Mitutoyo N0.2046 (0.01–10 mm, Kawasaki, Japan) analog micrometer gauge fitted with a spring-loaded elephant foot, and the mass measurements were performed with a Mettler Toledo XS204 Analytical Balance (Columbus, OH, USA) with a readability of 0.1 mg and a repeatability of 0.07 mg. Further details are shown in Supplementary Figure S2E–H. The electron microscopy imaging was performed with a field emission scanning electron microscope (TESCAN MAIA3-2016, Brno, Czech Republic) at 1 kV–3 kV with an SE detector. The top-view images are oriented so the lengthwise paper direction is upwards in the image. Simple light transmission tests were performed by placing the sample strips on a light table and studying the variance of transmitted light across the surface. Tape tests were used to study nanographite flake adhesion in the coating by placing and pulling Scotch tape strips on the top of the coating for each roll.

2.3.2. Lithium-Ion Anode Application

LIB half-cells were assembled as CR2025 coin cells under an argon atmosphere in a glove box. Discs of the coated paper with a 16 mm diameter were used as the anode and separator, wherein the paper side acted as a separator and the coating as an anode. Another 16 mm disc of copper foil was added as a current collector on the nanographite (anode) side, to allow comparison with traditional reference cells without influence from different electrode–contact interface materials. Three cells B3-1, B3-2, and B4-2 were fabricated with the Advantage Plus paper and from roll 03B. Two cells, B5-2, and B6-2, were made with Advantage Boost paper from roll 11. A 0.2 mm thick and 13 mm diameter disc of lithium metal foil from Goodfellow in England was used as a reference and counter-electrode (cathode). LP40 was used as an electrolyte (1M $LiPF_6$ in a mixture of ethylene carbonate and diethyl carbonate in a 1:1 weight ratio). Cyclic voltammetry (CV) tests of the cells were performed between 0.01 and 1.5 V at a scan rate of 0.1 mVs^{-1}. Galvanostatic charge–discharge (GCD) tests of the cells were performed at a current density of 100 mAg^{-1} in a voltage range between 1 mV and 1.0 V. The specific capacities and current densities were calculated based on the weight of the active materials of the electrode. All of the electrochemical measurements were conducted at room temperature. To study the ionic conductivity effects of the paper separator, an additional set of four coin cells, 4:1, 4:2, 4:3, and 4:4, were fabricated in a similar way but with additional uncoated papers inserted as extra separator materials, i.e., 4:1 contained one extra paper, 4:2 two and so on. The thickness of the uncoated paper sheets was 96.60(50) μm. These four cells were all made with Advantage Plus and roll 03B, and were then characterized for electric series resistance (ESR) performance with an Ametek Parstat potentiostat at both 1 mV DC (discharged state) and 1.5 V DC (charged state) during a traditional frequency sweep from 1 MHz to 100 mHz at a 10 mV AC amplitude. The resistivity of the paper separator was then calculated from the linear regression of ESR values from low-end frequency vs. total separator thickness. For comparison with the paper-based cells, a traditional reference cell was fabricated, cell R1, in which Slurry A was cast onto copper foil instead of paper at lab-scale. As a separator, a commercial Cellgard 2325 plastic membrane was used, and besides this, the rest of the assembly sequence as well as the electrolyte and lithium was the same as for the other paper-based cells.

3. Results

Table 1 shows a summary of the paper electrode rolls and their respective slurry, paper substrate, coat speed, transfer roll (TR) speed, metering nip, final coat weight, coating thickness, and calculated coating density.

The maximum coating thickness obtained during the stable coating operation with Slurry A was from roll 03B with a coat weight of 4.39(31) g/m^2 and a coating thickness of 16.4(1.0) μm. With higher coat weight, the coating operation became unstable, resulting in frequent paper failure, starting with wrinkling followed by paper breakage. Repeated coating on the same substrate paper (Plus) with Slurry B resulted in roll 06 and roll 07, differentiated by adjustment of the transfer roll speed. Roll 07 obtained a coating thickness of 46.8(2.1) μm at a coat weight of 11.42(31) g/m^2 during the stable coating operation. With identical settings as in roll 07, coating Slurry B on the Boost paper substrate resulted in roll 08 with a coat weight of 12.83(22) g/m^2 and a corresponding coating thickness of 25.9(1.1) μm. Roll 08 indicates that increased coat weight generates a higher density coating, suggesting a calendering effect of the increasing paper web tension when the coating faces a steel roll prior to winding. Further attempts were made to increase coat weight with maintained density by lowering the machine speed and transfer roll speed and thus allowing for decreased web tension; this resulted in roll 10 to roll 12. Roll 10 had insufficient drying and was still wet during winding.

The maximum coat weight achieved was 17.65(29) g/m^2 with Slurry B and Boost substrate paper for roll 11; the corresponding coating thickness was 47.7(1.4) µm. Roll 12 was made by repeating the setting of roll 06 but on the Boost substrate paper, generating a coat weight of 10.79(24) g/m^2 and a coating thickness of 56.1(2.5) µm. Rolls missing from the chronological numbering were either uncoated rolls used for paper samples or rolls made only to adjust the coating parameters to achieve stable coating operation. To obtain an indication of the influence of density on the physical and electrical properties, sheets cut from rolls with Slurry B were calendered without heat and included in the following results.

Table 1. Summary of the paper electrode rolls and main coating parameters. Values in brackets are the standard deviation error of the last two significant digits of the measurement values. Coating density is a calculated value based on measured thickness and weight.

Roll Number	Coating Slurry	Substrate Paper	Machine Speed [m/min]	TR Speed [m/min]	Metering Nip [mm]	Coat Weight [g/m^2]	Coating Thickness [µm]	Coating Density [g/cm^3]
UM	A	Plus	25	−700	35	4.07(28)	8.41(66)	0.48
03B	A	Plus	25	−900	38	4.39(31)	16.4(1.0)	0.27
06	B	Plus	25	−400	38	8.39(33)	25.4(1.1)	0.33
07	B	Plus	25	−800	38	11.42(31)	46.8(2.1)	0.24
08	B	Boost	25	−800	38	12.83(22)	25.9(1.1)	0.50
10	B	Boost	15	−500	28	13.19(75)	38.3(2.8)	0.34
11	B	Boost	20	−400	35	17.65(29)	47.7(1.4)	0.37
12	B	Boost	25	−400	38	10.79(24)	56.1(2.5)	0.19

3.1. Physical Properties

Figure 2 shows the coat weight as a function of coating thickness for Slurry A, Slurry B, and calendered samples from Slurry B, and the density of solid graphite as a reference.

Figure 3 shows a transmission light test for strips from each roll. A and C show the coating in ambient light conditions for strips in lengthwise and widthwise directions, respectively. B and C show the coating in transmitted light conditions for strips in lengthwise and widthwise directions, respectively. Samples 03A, 03B, 06, and 07 are on the Advantage Kraft Plus paper and 08, 10, 11 and 12 are on the Advantage Boost HP paper. In ambient light conditions, all lengthwise (A) samples, except roll 10, show a fairly even coating. Roll 10 has a spotted surface caused by insufficient drying and thus wet winding. In the widthwise samples, in ambient light conditions (C), a paper distortion feature can be seen. This feature is referred to as cockling and appears as ripples or wrinkles in areas with relatively high coat weight. Roll 08 and roll 11 have the least of this paper distortion. In transmitted light conditions, the quality of the coating can be seen. Roll 03B and UM in the lengthwise direction in transmitted light (B) conditions show clear strip patterns, while roll 08 and roll 11, with the thickest coatings, show a uniform coverage with almost no transmitted light passing through the coating.

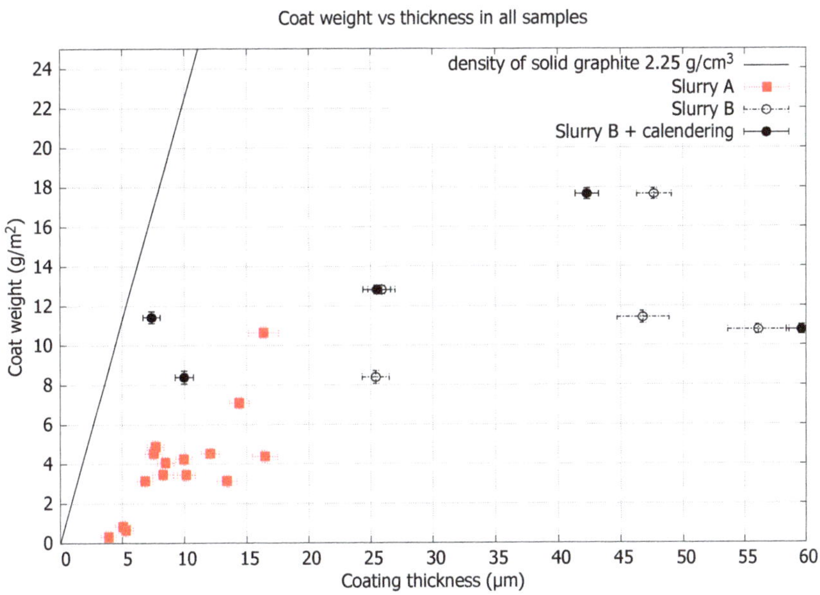

Figure 2. Coat weight as a function of coating thickness for Slurry A, Slurry B and calendered samples from Slurry B. The solid line represents the boundary when the density of coating becomes the same as that of solid graphite.

Figure 4A shows a tape test of the adhesion of the coating nanographite flakes. The top part of the image shows the pulled-off tape with a white background, and the lower part of the image shows the resulting mark on the strips from the tape for each roll. The tape tests show that only the top nanographite layer peels off with the tape and the adhesion appears to be similar for all samples. Figure 4B shows an SEM image on a cross-section of roll 11 where the red line marks the boundary between the coating (upper part) and the paper substrate (lower part). Roll 11 had the highest coat weight of all rolls, and the cross-section image shows a quite dense and mostly uniform coating. Figure 4C,D show SEM images of the surface of the uncoated Boost HP paper and the surface of the coated paper in roll 11, respectively. The uncoated Boost paper is a machine-finished paper, which is confirmed by the surface image (C) showing a smooth surface with compressed fibers; the image shown is the outside of the paper roll, i.e., the side that later was coated. From the surface image (D) of the coated paper, the coating appears thick with a smooth compressed surface and visual imprints caused by fibers on the paper backside during winding.

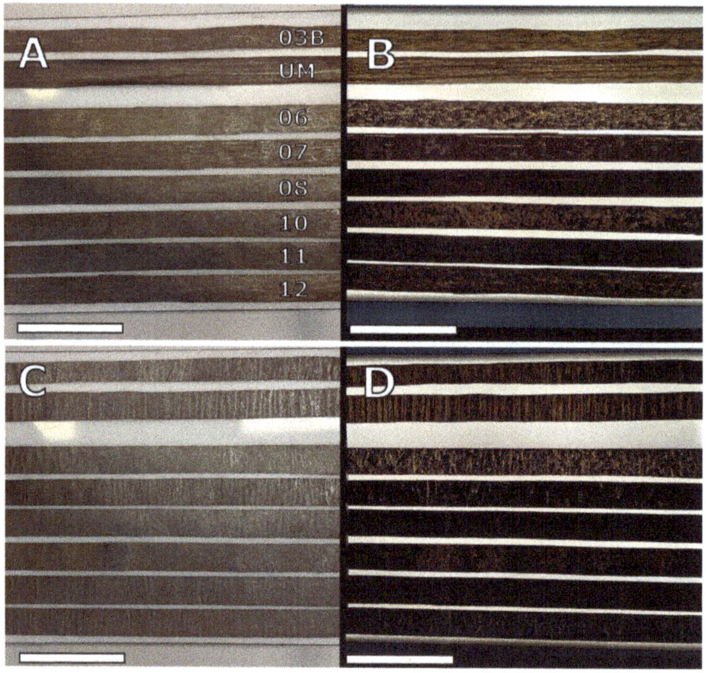

Figure 3. Transmission light test for strips from each roll. (**A**) shows the coating in ambient light conditions for strips in the lengthwise direction and in (**B**), the same strips are shown under transmitted light conditions. (**C**) shows the coating in ambient light conditions for strips in the widthwise direction, and in (**D**), the same strips are shown under transmitted light conditions. Scale bars are 10 cm.

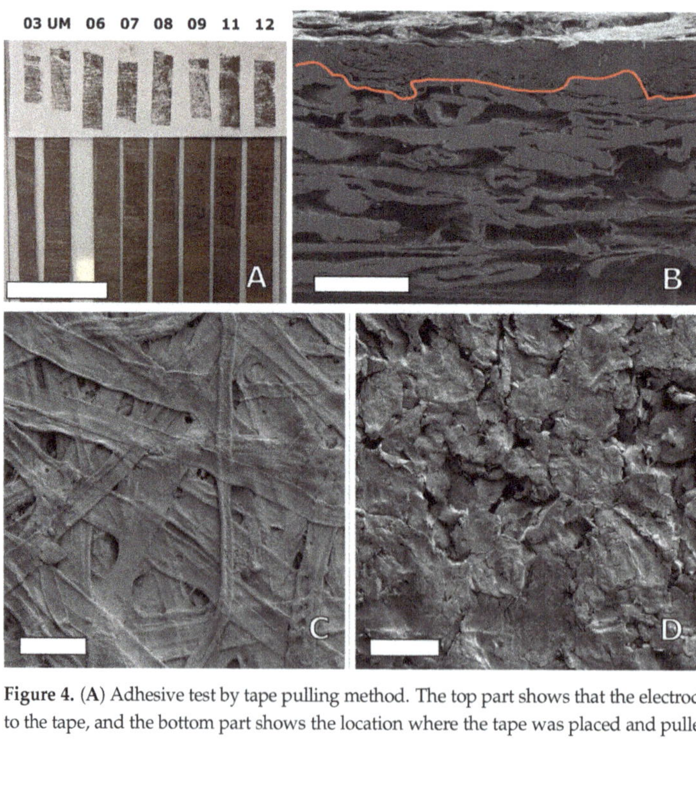

Figure 4. (**A**) Adhesive test by tape pulling method. The top part shows that the electrode remains stuck to the tape, and the bottom part shows the location where the tape was placed and pulled; the scale bar

is 10 cm. (**B**) SEM (SE) image of a cross-section from roll 11; the red line marks the boundary between the coating at the top and the paper substrate at the bottom; the scale bar is 50 µm. (**C**) SEM (SE) top-view image of uncoated Boost HP paper, and (**D**) top-view image of the coated roll 11; the scale bars are 100 µm in (**C**,**D**).

3.2. Electrical Properties

Figure 5 shows coating resistivity as a function of coating density for all samples made from Slurry A and Slurry B. Calendered samples of Slurry B are also included together with a graphite model representing the theoretical resistivity for solid graphite at different densities, assuming matching homogeneous porosity. From the diagram, it can be seen that the samples follow the same trend as the graphite model, where resistivity decreases with increasing coating density. The samples from Slurry A have relatively large variation, while the samples from Slurry B more clearly follow this trend.

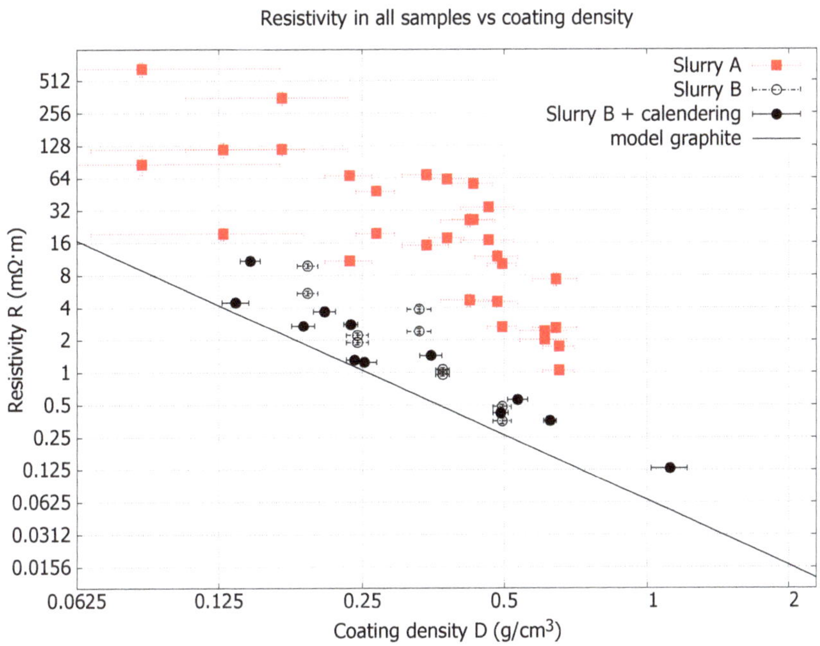

Figure 5. Coating resistivity as a function of coating density for all samples made from Slurry A and Slurry B. Calendering refers to sheets taken from coated rolls that were calendered without heat. The graphite model represents the theoretical resistivity for solid graphite at different porosities (densities). Error bars are standard deviation errors.

Figure 6 shows coating resistance as a function of distance from the paper edge in both lengthwise and widthwise directions from roll 06 and roll 11. Three strips were measured for each roll in the widthwise direction and one in the lengthwise direction. For both roll 06 and roll 11, each sample strip shows a linear increase in resistance with distance, indicating a uniform coating in both lengthwise and widthwise directions for both rolls. Resistivity was calculated via linear regression, and for roll 06, the resistivity in the lengthwise direction was L1 = 2.42(14) mΩ·m, and in the widthwise direction, it was W1 = 3.88(25) mΩ·m, W2 = 4.52(29) mΩ·m, and W3 = 3.81(17) mΩ·m. For roll 11, the resistivity in the lengthwise direction was L1 = 0.958(38) mΩ·m, and in the widthwise direction, it was W1 = 1.062(37) mΩ·m, W2 = 1.004(35) mΩ·m, and W3 = 0.999(37) mΩ·m. Roll 06 shows anisotropic behavior with a higher resistivity in the widthwise direction compared to the lengthwise direction, while roll 11 shows near-excellent isotropic behavior and also overall lower resistivity.

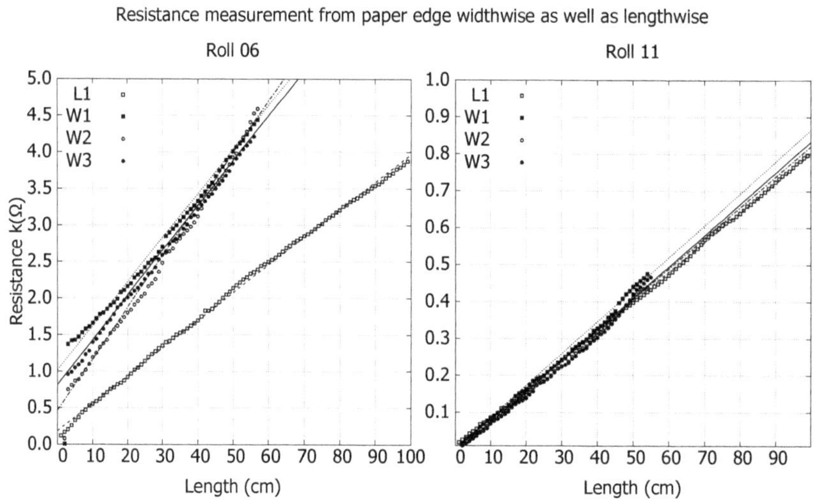

Figure 6. Resistance as a function of length from the edge on three strips cut in the widthwise direction and one strip in the lengthwise direction from roll 06 and roll 11. For roll 06: L1 = 2.42(14) mΩ·m, W1 = 3.88(25) mΩ·m, W2 = 4.52(29) mΩ·m, and W3 = 3.81(17) mΩ·m. For roll 11: L1 = 0.958(38) mΩ·m, W1 = 1.062(37) mΩ·m, W2 = 1.004(35) mΩ·m, and W3 = 0.999(37) mΩ·m.

3.3. LIB Anode Application

Roll 03B and roll 11 were further used as anodes for LIB coin cells in a half-cell configuration. Galvanostatic charge–discharge measurements of these cells were made to test the storage capacity of the electrode, whereas cyclic voltammetry measurements were made to study the reaction kinetics of the electrodes. Figure 7A shows a cyclic voltammogram for 100 cycles on a cell from roll 03B, and specific capacity from GCD as a function of cycle number is shown in Figure 7B. Cells B3-1, B3-2, B4-0, and B4-2 are identical, made with anodes from roll 03B, and have an electrode mass of 0.882(62) mg, while cells B5-2 and B6-2 are made from roll 11 with an electrode mass of 0.882(62) mg. From the cyclic voltammograms, it can be seen that both peak heights and area enclosed by the graph increase with the number of cycles. From GCD measurements, it can be seen that the specific capacity of the best-performing cell B3-2 increases linearly up to 100 cycles, reaching 147 mAh/g, and then remains stable until 300 cycles; this corresponds to 40% of the theoretical limit of 372 mAh/g of graphite electrode materials. After 300 cycles, the capacity slightly decreases with further cycling, and at 500 cycles, the capacity has decreased by 8.8%, corresponding to 134 mAh/g. Overall, the capacity for this cell remains fairly stable for extended cycling; the rest of the cells show unstable behavior. Figure 7C shows the corresponding cyclic voltammogram for the reference cell R1; the major anodic peak at 0.18 V is seen to be positioned to the left of the corresponding peak in the B4-0 cell at 0.24 V, a difference of 60 mV; a similar offset towards lower voltages is also seen for the other signatures on the cathodic side of the diagram as well. Figure 7D shows the results of the paper separator's influence and for the fully charged cell at 1.5 V; the resistivity of 241 Ωm is similar to the 237 Ωm found for an uncharged cell.

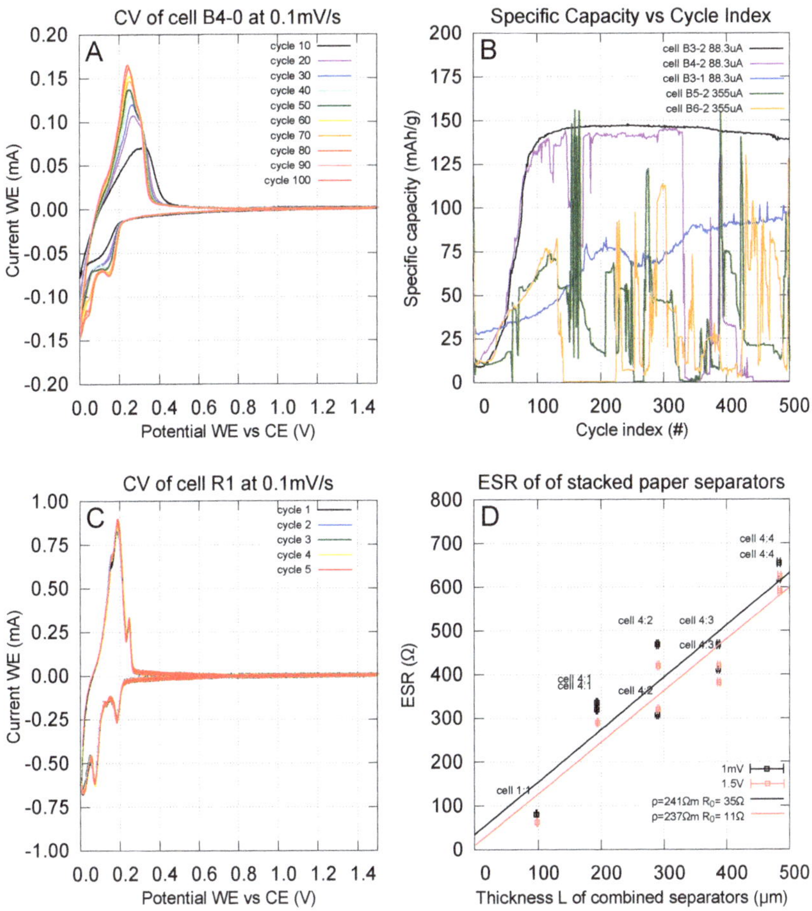

Figure 7. Cyclic voltammetry (**A**) on a half-cell configuration made from roll 03B and run for 100 cycles (35-day run time) at a rate of 0.1 mV/s. Galvanostatic charge–discharge cycling (**B**) for five half-cells; the total run time for cell B3-2 is 50 days for 500 cycles. Cells B3-1, B3-2, B4-0, and B4-2 are all identical and made with electrodes from roll 03B, corresponding to an electrode mass of 0.882(62) mg, while cells B5-2 and B6-2 are made from roll 11, mass 0.882(62) mg. Cyclic voltammetry (**C**) on reference cell R1, a graphite-coated copper foil instead of paper. Comparison (**D**) of electric series resistance (ESR) conducted by inserting additional sheets of paper as extra separator material, each 96.60(50) μm, showing that the resistivity of the Advantage Kraft Plus paper is 241 Ωm for a cell in a fully charged state.

4. Discussion

4.1. Physical Properties

The stability of operations was improved with the Boost paper substrate, and rolls 08 and 11 show no obvious wrinkling or "cockling"; see Supplementary Figure S2C,D. The physical appearance of the coated surfaces was also different between Slurries A and B when observed by SEM; this can be linked to the two different graphite sources as well as the exfoliation procedures used. It was found that Slurry A has smaller flake size dimensions compared to Slurry B, by about ten times; see Supplementary Figure S6 for further details. Regarding the evenness of coating, striping is seen in Slurry A samples, i.e., roll 03B and UM; visually, it is a line pattern in the traveling direction of the paper during coating, which is an artifact from the LAS coating technology. With better parameter settings and increased coating thickness, these lines disappear and coating becomes more uniform overall, and this is seen for the other rolls of Slurry B; see Figure 3D. In the same figure, one can also see that the coating thickness of roll 06 is not sufficient and the coating looks largely uneven; this is caused by the coating technology in that the valleys of the paper surface are filled first, while the high mountain point, i.e., on top of the outmost positioned fibers in the paper, remains exposed if the coating thickness is too thin. This LAS coating method gives the option of large variations in thickness; an order of magnitude span was achieved during these trials, and when complemented with an extra calendering post-processing step, the density of the coating was increased to 1.12 g/cm^3 and was very close to solid graphite levels, i.e., 2.25 g/cm^3, as seen in Figure 2. For comparison, standard LIB 18,650 cells have a coating thickness typically between 30 μm and 90 μm and an electrode density around 1 g/cm^3 [28]. In roll 11, the highest achieved coating thickness was 47.7(1.4) μm with a coat weight of 17.65(29) g/cm^2, indicating that the coating technique allows for sufficient coating thicknesses. Considering that coat weight is strongly dependent on the slurry's solids content, it is likely possible to also achieve equivalent sufficient coat weights with an improved slurry and further optimized coating operations. Besides the slurry itself, the combination of coating parameters together with the substrate properties is crucial to reach a specific electrode thickness or density. The amount of wet material (slurry) that is applied to the paper substrate can be varied by adjusting the machine speed, transfer roll speed, and metering nip, but the paper properties (e.g., absorption rate and wet expansion) affect the web tension in the coating process, which affects the calendering effect (compression) on the electrode from the rolls before winding. The drying conditions are not assumed to have the same effect on electrode density but still need to be adjusted according to the wet coat weight to ensure that sufficient drying occurs without overheating the paper or slurry. Electrode density is important for electrodes used in battery applications, where a compact electrode for small ions is preferred. Hence, this coating method allows for tailoring the electrode density for different types of application. The LAS coating technology used is directly scalable to industrial levels. LAS can be scaled up to paper widths exceeding 6 m for large-scale production, and the maximum machine speed is above 2000 m/min. In a possible maximum scale-up for roll 11, the machine speed could be increased by a factor of 7.5 (limited by the speed of the transfer roll), and with a 6 m coating width, the production rate would reach 900 m^2/min in a single production line. In the traditional industrial LIB cell production process, the coat speed is in the range of 35 m/min to 80 m/min with a coating width of up to 1.5 m [29]. To the best of our knowledge, no study has been published on the large pilot-scale coating of paper for electrode applications. However, for comparison of the achieved scale-up performances of our work and to visualize the advantages in the speed of area production by utilizing coating methods on top of existing substrates, it can be clearly seen that the small pilot trial that demonstrated a paper conversion approach and was conducted at a web width of 20 cm and a speed of 1 m/min is slower [19]. For a more proper comparison with similar materials, the spray-coating approach in a paper fabrication process at a large pilot scale achieves results in the same range but somewhat slower with 15 m/min, a similar web width, and a lower electrode/spray weight of 11.6 g m^2, although how a further scale-up of this spray method from pilot to industry with a preserved consistent quality over web widths is supposed to be carried out is uncertain [22].

4.2. Electrical Properties

Anisotropic behavior of resistivity is favorable in some applications, for example, in thermal applications of conduction of heat since heat transfer is linked to electrical transport phenomena; however, in this study, the focus was on achieving isotropic behavior, which was also demonstrated for the thicker coatings like roll 11, as shown in Figure 6. The threshold for isotropic behavior is roll 07, i.e., a coat weight of 11.42 g/m^2; all samples with a lower coat weight than that showed anisotropic behavior. This anisotropic behavior can be explained by two things: First, the LAS coating technique can influence it, as seen in the Slurry A trials, and if not adjusted properly, when coatings are thin, it forms strip patterns across the width of the roll, leading to very high anisotropic behavior in resistivity. Secondly, the anisotropy of the paper favors the fibers aligning in the lengthwise direction during paper manufacturing, leading to a smoother surface and hence, for a thin coating, keeping an electrical connection between the nanographite flakes is easier compared to going across fibers in the widthwise direction. For more details on this, see Supplementary Figures S7 and S8.

It is worth mentioning that the method used for measuring the resistivity also provides an excellent way to characterize the overall coating quality; thus, when making resistance vs. length measurements, one can clearly see if there is a coating defect and, for example, if the coat weight changes across the width of the paper, which can happen with improper setting of the LAS system; this is seen in roll 06 in the left graph of Figure 6 for the W1 strip, showing a nonlinear trend with a decreased slope in the data curve towards the edge of paper, indicating that coat weight increased by about 1/3.

In Figure 5, two results can be observed. Firstly, the best performance was found in the calendered sheets, giving values close (50%) to the theoretical limit of porous graphite structures with similar densities to the coating. This indicates that the connections between the nanographite flakes are overall good and are not hampered by the added dispersive agent during slurry mixing. Secondly, Slurry B outperforms Slurry A in that the former is in all coating trials closer to the theoretical limit of graphite. This could perhaps be explained by the difference in nanographite flake size and the fact that Slurry A has smaller sizes compared to Slurry B, and hence, hypothetically, if the same coating density is achieved in two coatings but different slurries are used, then Slurry A will have more interconnected electrical resistance points between the flakes since the number of flakes becomes larger to compensates for their smaller sizes; see Supplementary Figure S6A,B for detailed images of flake morphology. Overall, a correlation study like Figure 5 is a good method to characterize the material. Since the resistivity dependency of density says something about the internal structure of the material and since we have two flake size distributions in our experiments, we can see that this is also observed as a feature in that same figure; a future investigation could be to correlate this resistivity change with the different dispersive agents used.

4.3. LIB Anode Application

The cyclic voltammograms (Figure 7A) show an increase in the peak height of the CV curve during the cycling. A higher peak indicates more lithium ions being inserted into the anode. Higher peaks suggest that a larger portion of the anode material actively participates in the electrochemical reaction overall. However, it is observed that for the paper electrode, this process of activating the electrode is rather slow compared to more standard cells, and we do not see a stabilization in its performance until 100 cycles.

The paper electrode has all its peaks offset to higher voltage values compared to the reference cell R1's CV data (Figure 7C). For the highest peak on the anodic side, this offset is about 60 mV; this is comparable to the influence of the paper separator's voltage contribution due to added series resistance. Measured (Figure 7D) with a resistivity of 241 Ωm, a single paper layer as used for cell B4-0 will correspond to an added series resistance of 116 Ω, and at the peak current of 165 μA, this will add a voltage offset of 19.1 mV. If compared with R1, one has to also take into account the voltage contribution of the

Celgard membrane as well, but this has a much lower value than the paper's contribution, an estimate since the membrane is 25 μm in thickness, which is four times thinner than the paper, and with the approximation and all else being equal, this contribution is about 4 mV and the offset between the two cells caused by separator influences will be 15 mV. Hence, the observed offset could be explained as being at least partly caused by the paper's higher electrical resistance.

The long-term stability of battery capacity over extended cycling highlights the potential of using paper as a separator. Furthermore, this also strengthens the potential of a new design concept wherein paper can act as both a separator and a substrate for coating the anode material. A unique advantage of this approach is that the electrode is electrochemically accessible from both of its sides since the graphite in this case is not coated onto a metallic foil, which is the more traditional approach.

Other Applications of the Coated Paper Electrode

The early coating experiments prior to the content of this study have shown that triboelectric nanogenerators utilizing coated electrode paper from another batch of Slurry B on Kraft Plus paper can generate a peak power density of 14 kW/m^2, driven by the electrostatic discharge effect on the material's surface [25]. The same study also showed that the coated paper could be used for wall- or floor-mounted smart sensors to detect movement as well as smart sheets that monitor body movements and physiological activities during sleep. Furthermore, lab-scale coating experiments prior to this study have shown that other slurry batches made with the same equipment coated on paper can also be used as electrodes in aqueous, metal-free, and low-cost supercapacitors generating a specific capacitance of 48 F/g [30].

5. Conclusions

This study revealed the following two conclusions. First, it was demonstrated that a functional electrically conducting material coated onto a paper substrate could be produced at a pilot scale (TRL6) with roll-to-roll methods and that this material shows promising performance for further electrode applications, especially since it is environmentally friendly, resource-efficient, and low-cost. The best values achieved in this study were a coat weight of 17.65(29) g/m^2 (roll 11), a max coating thickness of 56.1(2.5) μm (roll 12), the highest density of 1.117(97) g/cm^3 (roll 08), and the highest electrical conductivity of 0.1293(17) mΩ·m (roll 08), making roll 08 the best-performing coating on average. Secondly, the two characterization methods used to measure resistance as a function of a dimensional distance and study resistivity as a function of density are both good tools for further materials science investigations when characterizing fundamental properties of new materials at the nanoscale.

Supplementary Materials: The following supporting information can be downloaded at https://www.mdpi.com/article/10.3390/nano15020113/s1, Figure S1: (A) shows the high pressure piston pump connected to the helical coil tube shear zone. (B) shows the two processing containers, one feeding the the exfoliation process with suspension and one retreving exfoliated material. (C) shows the shear zone outlet in container 2 during the second full pass; Figure S2: (A,B) shows the coated paper substrate exiting the coater applicator and entering the IR-drying section from two different angles. (C) shows sample cut-outs from Roll 08 and (D) from Roll 11. (E) shows 3 cm sample strips cut across the full with from one of the rolls, (F) the procedure och punching 16 mm coins from each sample strip, (G) shows weight measurement of punched out sample disks, (H) shows thickness measurement of the sample discs and (I) a schematic sketch of the LIB half-cell assembly; Figure S3: Humidity dependency tested for coated sheets 02B (top) and 02C (bottom). Both sheets 02B and 02C was made with Slurry A on Plus paper substrate. Precut A4 sample sheets from coating trials 02B and 02C was cut into five stripes 10 mm wide, 30 cm long and with guillotine. Room humidity was measured to be 22%RH followed by electrical resistance measurement with multimeter and this by placing the stripes on a plastic ruler with the electric connector-clamps at its ends. Each stripe were thereafter put into desiccator boxes with salt-humidifiers giving a fixed humidity of 65%RH, 69%RH, 72%RH, 75%RH, and 84%RH for 24 h periods with followup resistance measurements. All samples were then in 24 h intervals rotated between all the humidity options giving five data sets for each 02B and 02C sample. Experiment ended with a final treatment of all samples in oven to be dried over night at 69C. Results shows very small trend of increasing resistance with humidity but is insignificant if compared to the error estimates of data, (error bars only shown for top row of dataset for better clarity in graph); Figure S4: Electrical resistance was measured with multimeter in centimeter intervals over the length of stripes that were cut into lengthwise sections taken from roll 03B and UM. The measurement was repeated after 7–8 mounts passed. For roll 03B the resistivity increased by a minor 22% and for roll 11 no change was observed; Table S1: Two sheets ofA4 size were cut from each of the coated rolls, one sheet was calendered without heat, while the second sheet was left for comparison. Stripes about one inch width was cut out from the sheets in lengthwise direction of paper followed by thickness-, width-, and electrical-resistance measurements in similar way as described in method section. Calculation of coating compression was done as well as comparison of coating thickness reduction, density increase, and electrical resistivity decrease due to the calendering. Calendering effects were not studied on roll 03B and UM. * referees to uncoated papers hence weight, thickness, density, and compression ratio is for the full paper. † for roll 10 the coating was too uneven and when two samples were compared the individual variation between samples was larger than the effect of the calendering leading to a results in which the calendering sample appeared thicker than the uncalender version; Figure S5: Comparison of resistivity decrease as function of compression of the coating due to calendering. Sheet 06 to 11 follows a linear behavior as expected for different compression ratios. However the change is not 1:1, i.e., if the compression is 60% as for sheet 06 a change in resistivity should also be 60% since resistivity depends linearly on thickness of coating but this is not observed, instead only 40% resistivity decrease is found and the ratio becomes 1.37 between resistivity decrease and compression. The regression gives an intercept of y-axis at −41.7 with is unrealistic. Sheet 12 was excluded from the regression calculations and is seen far of from the trend line, this can be explained by the achieved lower coating density of Sheet 12 is not caused by a uniform porosity increase but rather of uneven coating which is also confirmed in SEM imaging; Figure S6: **A** sample, **B** machine-speed (m/min), **C** transferroll speed (m(min), **D** coating thickness (um), **De** thickness error (um), **E** coat weight (g/m^2), **Ee** coat weight error (g/m^2), **F** ρ widthwise (mΩm), **Fe** ρ w.error (mΩm), **G** ρ lengthwise (mΩm), **Ge** ρ l.error (mΩm), **H** ratio ρ width/length, **He** ratio ρ error, **I** measured sheet resistance (kΩ/sq), **Ie** measured sheet resistance error (kΩ/sq), **J** calculated sheet resistance (kΩ/sq), **Je** calculated sheet resistance error (kΩ/sq), **K** measured capacitance (F/g), **Ke** measured capacitance error (F/g). For sheet resistance two methods was used, either (column **I**) direct measurement by four point probe pushed against coating surface at one hundred locations for each sample or calculated from the bulk resistivity ρ (column **J**). All errors are std. dev. errors; Figure S6: Top row: SEM images from roll 03B (A) detailed top view, (B) lower magnification of top view. Bottom row in similar way but for roll 11, (C) detailed top view, (D) lower magification of top view. Scalebars from left to right both rows: 100 µm and 500 µm. Except for the thinner coating on roll 03B compared to roll 11 the other observations seen are differences in flake size where approximately 10 times smaller flakes are present in slurry A, as seen in the detailed image (A) of roll 03B, compared to (C) made from Slurry B. Another observation is that paper fibers can be seen through the coating at some places (B) compared to (D); Figure S7: SEM crossectional image in paper lengthwise direction, (A) from roll 03B and (B) from roll 11, scalebars 200 µm. For 03B the coating looks thinner in that the top fibers of the paper are not coated and only the valleys of the paper surface are filled with the nanographite. It also looks somewhat more porous compared to roll 11. Since there is a natural fiber anisotropy of a paper in the favoring alignment of fibers in paper length direction a thin coating that only fills the voids of the paper therefore tends to connect lengthwise at first hand and secondly orthogonal across the fibers, hence leading to anisotropy effects of electrical resistivity with higher resistivity in the width direction of the paper rolls; Figure S8: SEM crossectional image in sheets from 02C roll. Image (A) corresponds to crossection taken in the width direction of the roll and (B) the length direction. Scalebars 200 μm. As seen in (A) the surface of the paper is more rough in the horizontal direction of the image compared to image (B) indicating the expected anisotropy of the paper; Figure S9: High resolution SEM crossectional image, scalebar 20 μm, from roll 07 showing the nanographite coating layer on top of the paper. Here thickness of coating is measured to be 32.8 μm and the nanographite flakes tend to aggregate into bundles and these are about 3.1 μm in thickness. The thinnest nanographite sections seen in this image is 100 nm. The big solid region at the bottom of image is the top fiber of the paper substrate; Figure S10: SEM images of crossection as well as the top of coating in perspective, scalebars 500 μm. Image (A) shows roll 12 and the dark regions shows regions with little and in top left of image no coating and with exposed fibers, indicated by the observed charging artifacts in the SEM image. This is seen as one cause of the different behavior of roll 12 in the resistivity measurements in that the lower density coating is not homogenies but rather caused by these larger variations in the coating. Image (B) shows the other end of the spectra the smoothest surface achieved during the pilot trials and is from roll 11 after calendering.

Author Contributions: Conceptualization, N.B. and M.H.; data curation, M.H.; formal analysis, N.B., M.P. and M.H.; funding acquisition, N.B.; investigation, N.B., M.P., R.P., R.Z., N.L. and M.H.; methodology, N.B., M.P. and M.H.; project administration, N.B.; resources, N.B. and M.H.; supervision, N.B.; validation, N.B., M.P., R.P., R.Z. and M.H.; visualization, N.B. and M.H.; writing—original draft, N.B. and M.H.; writing—review and editing, N.B., M.P., R.P., N.L. and M.H. All authors have read and agreed to the published version of the manuscript.

Funding: This research was funded by the SIO Grafen programme funded by Vinnova, the Swedish Energy agency and Formas (reference number: 2021–04083).

Data Availability Statement: Data are contained within the article or Supplementary Materials.

Acknowledgments: The authors would like to acknowledge the financial support within the SIO Grafen programme funded by Vinnova, the Swedish Energy agency and Formas (reference number: 2021–04083) as well as the participating project partners UMV coating systems AB and Mondi Dynäs AB.

Conflicts of Interest: The authors declare no conflicts of interest and the funders had no role in the design of the study; in the collection, analyses, or interpretation of data; in the writing of the manuscript; or in the decision to publish the results.

References

1. Zhang, Y.S.; Schneider, K.; Qiu, H.; Zhu, H.L. A perspective of low carbon lithium-ion battery recycling technology. *Carbon Capture Sci. Technol.* **2022**, *5*, 100074. [CrossRef]
2. Xu, P.; Tan, D.H.; Chen, Z. Emerging trends in sustainable battery chemistries. *Trends Chem.* **2021**, *3*, 620–630. [CrossRef]
3. Jiang, S.Q.; Nie, C.C.; Li, X.G.; Shi, S.X.; Gao, Q.; Wang, Y.S.; Zhu, X.N.; Wang, Z. Review on comprehensive recycling of spent lithium-ion batteries: A full component utilization process for green and sustainable production. *Sep. Purif. Technol.* **2023**, *315*, 123684. [CrossRef]
4. Huang, B.; Pan, Z.; Su, X.; An, L. Recycling of lithium-ion batteries: Recent advances and perspectives. *J. Power Sources* **2018**, *399*, 274–286. [CrossRef]
5. Zhao, Y.; Yuan, X.; Jiang, L.; Wen, J.; Wang, H.; Guan, R.; Zhang, J.; Zeng, G. Regeneration and reutilization of cathode materials from spent lithium-ion batteries. *Chem. Eng. J.* **2020**, *383*, 123089. [CrossRef]
6. Fan, E.; Li, L.; Wang, Z.; Lin, J.; Huang, Y.; Yao, Y.; Chen, R.; Wu, F. Sustainable recycling technology for Li-ion batteries and beyond: challenges and future prospects. *Chem. Rev.* **2020**, *120*, 7020–7063. [CrossRef]
7. Yu, X.; Manthiram, A. Sustainable battery materials for next-generation electrical energy storage. *Adv. Energy Sustain. Res.* **2021**, *2*, 2000102. [CrossRef]
8. Al Rai, A.; Yanilmaz, M. High-performance nanostructured bio-based carbon electrodes for energy storage applications. *Cellulose* **2021**, *28*, 5169–5218. [CrossRef]
9. Liedel, C. Sustainable battery materials from biomass. *ChemSusChem* **2020**, *13*, 2110–2141. [CrossRef] [PubMed]

10. Cao, S.; Shi, L.; Miao, M.; Fang, J.; Zhao, H.; Feng, X. Solution-processed flexible paper-electrode for lithium-ion batteries based on MoS2 nanosheets exfoliated with cellulose nanofibrils. *Electrochim. Acta* **2019**, *298*, 22–30. [CrossRef]
11. Devi, M.; Wang, H.; Moon, S.; Sharma, S.; Strauss, V. Laser-Carbonization—A powerful tool for micro-fabrication of patterned electronic carbons. *Adv. Mater.* **2023**, *35*, 2211054. [CrossRef]
12. Lee, M.H.; Lee, J.; Jung, S.K.; Kang, D.; Park, M.S.; Cha, G.D.; Cho, K.W.; Song, J.H.; Moon, S.; Yun, Y.S.; et al. A biodegradable secondary battery and its biodegradation mechanism for eco-friendly energy-storage systems. *Adv. Mater.* **2021**, *33*, 2004902. [CrossRef] [PubMed]
13. Roy, J.J.; Zaiden, N.; Do, M.P.; Cao, B.; Srinivasan, M. Microbial recycling of lithium-ion batteries: Challenges and outlook. *Joule* **2023**, *7*, 450–456. [CrossRef]
14. Nguyen, T.H.; Fraiwan, A.; Choi, S. Paper-based batteries: A review. *Biosens. Bioelectron.* **2014**, *54*, 640–649. [CrossRef] [PubMed]
15. Juqu, T.; Willenberg, S.C.; Pokpas, K.; Ross, N. Advances in paper-based battery research for biodegradable energy storage. *Adv. Sens. Energy Mater.* **2022**, *1*, 100037. [CrossRef]
16. Zolin, L.; Destro, M.; Chaussy, D.; Penazzi, N.; Gerbaldi, C.; Beneventi, D. Aqueous processing of paper separators by filtration dewatering: towards Li-ion paper batteries. *J. Mater. Chem. A* **2015**, *3*, 14894–14901. [CrossRef]
17. Zahn, R.; Lagadec, M.F.; Hess, M.; Wood, V. Improving ionic conductivity and lithium-ion transference number in lithium-ion battery separators. *ACS Appl. Mater. Interfaces* **2016**, *8*, 32637–32642. [CrossRef]
18. Brooke, R.; Åhlin, J.; Hübscher, K.; Hagel, O.; Strandberg, J.; Sawatdee, A.; Edberg, J. Large-scale paper supercapacitors on demand. *J. Energy Storage* **2022**, *50*, 104191. [CrossRef]
19. Sandberg, M.; Tordera, D.; Granberg, H.; Sawatdee, A.; Dedic, D.; Berggren, M.; Jonsson, M.P. Photoconductive zinc oxide-composite paper by pilot paper machine manufacturing. *Flex. Print. Electron.* **2016**, *1*, 044003. [CrossRef]
20. Yao, B.; Zhang, J.; Kou, T.; Song, Y.; Liu, T.; Li, Y. Paper-based electrodes for flexible energy storage devices. *Adv. Sci.* **2017**, *4*, 1700107. [CrossRef] [PubMed]
21. Hu, L.; Choi, J.W.; Yang, Y.; Jeong, S.; La Mantia, F.; Cui, L.F.; Cui, Y. Highly conductive paper for energy-storage devices. *Proc. Natl. Acad. Sci. USA* **2009**, *106*, 21490–21494. [CrossRef] [PubMed]
22. Beneventi, D.; Chaussy, D.; Curtil, D.; Zolin, L.; Bruno, E.; Bongiovanni, R.; Destro, M.; Gerbaldi, C.; Penazzi, N.; Tapin-Lingua, S. Pilot-scale elaboration of graphite/microfibrillated cellulose anodes for Li-ion batteries by spray deposition on a forming paper sheet. *Chem. Eng. J.* **2014**, *243*, 372–379. [CrossRef]
23. Zolin, L.; Destro, M.; Curtil, D.; Chaussy, D.; Penazzi, N.; Beneventi, D.; Gerbaldi, C. Flexible cellulose-based electrodes: Towards eco-friendly all-paper batteries. *Chem. Eng. Trans.* **2014**, *41*, 361–366.
24. Yang, P.; Li, J.; Lee, S.W.; Fan, H.J. Printed zinc paper batteries. *Adv. Sci.* **2022**, *9*, 2103894. [CrossRef] [PubMed]
25. Zhang, R.; Chen, D.; Hummelgård, M.; Blomquist, N.; Dahlström, C.; Chen, W.; Li, J.; Örtegren, J.; Wang, Z.L. Engineering Triboelectric Paper for Energy Harvesting and Smart Sensing. *Adv. Mater.* **2024**, 2416641. . [CrossRef] [PubMed]
26. Blomquist, N.; Engström, A.C.; Hummelgård, M.; Andres, B.; Forsberg, S.; Olin, H. Large-scale production of nanographite by tube-shear exfoliation in water. *PLoS ONE* **2016**, *11*, e0154686. [CrossRef]
27. Blomquist, N.; Alimadadi, M.; Hummelgård, M.; Dahlström, C.; Olsen, M.; Olin, H. Effects of geometry on large-scale tube-shear exfoliation of graphite to multilayer graphene and nanographite in water. *Sci. Rep.* **2019**, *9*, 8966. [CrossRef]
28. Bischof, K.; Marangon, V.; Kasper, M.; Regalado, A.A.; Wohlfahrt-Mehrens, M.; Hölzle, M.; Bresser, D.; Waldmann, T. Evaluation of commercial 18,650 and 26,700 sodium-ion cells and comparison with well-established lithium-ion cells. *J. Power Sources Adv.* **2024**, *27*, 100148. [CrossRef]
29. Heimes, H.H.; Kampker, A.; Lienemann, C.; Locke, M.; Offermanns, C.; Michaelis, S.; Rahimzei, E. *Lithium-Ion Battery Cell Production Process*; PEM der RWTH Aachen University: Aken, Germany, 2018.
30. Blomquist, N.; Koppolu, R.; Dahlström, C.; Toivakka, M.; Olin, H. Influence of substrate in roll-to-roll coated nanographite electrodes for metal-free supercapacitors. *Sci. Rep.* **2020**, *10*, 5282. [CrossRef] [PubMed]

nanomaterials

MDPI

Article

$Na_3MnTi(PO_4)_3$/C Nanofiber Free-Standing Electrode for Long-Cycling-Life Sodium-Ion Batteries

Debora Maria Conti [1], Claudia Urru [1], Giovanna Bruni [1], Pietro Galinetto [2], Benedetta Albini [2], Vittorio Berbenni [1], Alessandro Girella [1] and Doretta Capsoni [1,*]

[1] Department of Chemistry, Physical Chemistry Section & C.S.G.I. (Consorzio Interuniversitario per lo Sviluppo dei Sistemi a Grande Interfase), University of Pavia, Via Taramelli 16, 27100 Pavia, Italy; deboramaria.conti01@universitadipavia.it (D.M.C.); claudia.urru01@universitadipavia.it (C.U.); giovanna.bruni@unipv.it (G.B.); vittorio.berbenni@unipv.it (V.B.); alessandro.girella@unipv.it (A.G.)

[2] Department of Physics, University of Pavia, Via Bassi 6, 27100 Pavia, Italy; pietro.galinetto@unipv.it (P.G.); benedetta.albini@unipv.it (B.A.)

* Correspondence: doretta.capsoni@unipv.it; Tel.: +39-0382-987213

Abstract: Self-standing $Na_3MnTi(PO_4)_3$/carbon nanofiber (CNF) electrodes are successfully synthesized by electrospinning. A pre-synthesized $Na_3MnTi(PO_4)_3$ is dispersed in a polymeric solution, and the electrospun product is heat-treated at 750 °C in nitrogen flow to obtain active material/CNF electrodes. The active material loading is 10 wt%. SEM, TEM, and EDS analyses demonstrate that the $Na_3MnTi(PO_4)_3$ particles are homogeneously spread into and within CNFs. The loaded $Na_3MnTi(PO_4)_3$ displays the NASICON structure; compared to the pre-synthesized material, the higher sintering temperature (750 °C) used to obtain conductive CNFs leads to cell shrinkage along the *a* axis. The electrochemical performances are appealing compared to a tape-casted electrode appositely prepared. The self-standing electrode displays an initial discharge capacity of 124.38 mAh/g at 0.05C, completely recovered after cycling at an increasing C-rate and a coulombic efficiency ≥98%. The capacity value at 20C is 77.60 mAh/g, and the self-standing electrode exhibits good cycling performance and a capacity retention of 59.6% after 1000 cycles at 1C. Specific capacities of 33.6, 22.6, and 17.3 mAh/g are obtained by further cycling at 5C, 10C, and 20C, and the initial capacity is completely recovered after 1350 cycles. The promising capacity values and cycling performance are due to the easy electrolyte diffusion and contact with the active material, offered by the porous nature of non-woven nanofibers.

Keywords: Na-ion batteries; $Na_3MnTi(PO_4)_3$; NASICON-type electrode; self-standing electrode; carbon nanofibers

1. Introduction

The increasing world energy demand and the need to face global climate changes and environmental pollution concerns urgently call for the development of renewable energy sources such as solar, water, and wind [1–3]. Rechargeable batteries play a relevant role in large-scale renewable and clean energy storage [4–6]. Despite the fact that lithium-ion batteries (LIBs) exhibit a high operating voltage, high energy density, and improved specific capacity and lifespan [7–11] and represent a mature and performant technology in energy storage systems, the scarce and uneven distribution of lithium resources, as well as the rising cost, make them not suitable to fulfil large-scale applications [5,12–14]. Sodium-ion batteries (SIBs) represent a fascinating candidate for next-generation large-scale grid energy systems: SIBs' working mechanism is comparable to LIBs'. Sodium is abundant on the Earth's crust and seawater, and it is inexpensive [15–17]. Compared to lithium, sodium displays similar chemical, electrochemical, and physical properties, but its larger cation size (1.02 Å) and slower diffusion rate cause huge volume expansions during its insertion into the electrode material. This feature poses concerns on the SIBs' requisites such as adequate values of specific capacities, a sufficient coulombic efficiency, and long lifespan. The cathode materials play a relevant role in SIBs, and numerous compounds displaying open framework structures and suitable tunnels for Na-ion diffusion have been investigated, such as Prussian blue analogues [18–23], O3- and P2-type transition metal layered oxides [24–28], and polyanionic compounds [29–34]. Among the polyanionic phosphates, Na superionic conductor (NASICON)-structured compounds are appealing candidates as cathodes for SIBs. $Na_3V_2(PO_4)_3$ is considered the most representative: the Na ions partially occupy two crystallographic sites, and cation transport is assured by the open three-dimensional structure based on corner-sharing VO_6 octahedra and PO_4 tetrahedra to form $V_2(PO_4)_3$ repeating units [35]. More recently, research has focused on NASICON-structured mixed transition metal phosphates, suitable to access high voltages. Among them, $Na_3MnTi(PO_4)_3$ seems very promising. Gao and co-workers [36,37] detected two high-voltage plateaus at about 3.6 and 4.1 V, attributed to the Mn^{3+}/Mn^{2+} and Mn^{4+}/Mn^{3+} redox processes (the extraction of two sodium ions). After cycling and repeated intercalation/deintercalation processes, small volume changes and structural stability were detected. The Zhu research group [38] synthesized, through the spray-drying-assisted route, $Na_3MnTi(PO_4)_3$/C hollow microsphere cathodes, which exhibited a fully reversible three-sodium-ion extraction/insertion. The peaks at about 2.1, 3.5 and 4.0 V vs. Na^+/Na correspond to the Ti^{3+}/Ti^{4+}, Mn^{2+}/Mn^{3+}, and Mn^{3+}/Mn^{4+} redox reactions. A high specific capacity of 160 mAh/g at 0.2 C is achieved, comparable to the theoretical one (176 mAh/g). Despite the intriguing electrochemical properties of the $Na_3MnTi(PO_4)_3$ cathode, its poor electronic conductivity and the presence of intrinsic anti-site defects involving the Mn occupation of vacancies on Na(2) site cause voltage hysteresis and inhibit good cycling stability and rate performance [39]. Several strategies were investigated to overcome these problems: carbon coating, the synthesis of tailored morphologies/nanostructures [40,41], and the doping/partial substitution of transition metal ions [39,42–45].

In the last decade, carbon nanofibers (CNFs) proved to be appealing materials for electrochemical applications. The good electronic conductivity and mechanical properties make them suitable conductive fillers or conductive supports of electrode active materials [46–51]. Noteworthy, the CNFs themselves are promising anodes [52]: CNFs display a high surface area and electronic conduction thanks to their unique 1D morphology. They can be obtained by electrospinning, a simple and scalable technology: a polymeric solution is electrospun, stabilized at about 250 °C in air, and carbonized at 800–900 °C in an inert atmosphere. The obtained non-woven CNFs display good electronic conductivity and desirable porosity to allow electrolyte permeation and to buffer the volume changes occurring with sodiation/desodiation. Notably, the mechanical properties of CNFs make them suitable to fabricate self-standing electrodes, avoiding the use of the metal current collector. Two approaches are mainly used to synthesize active materials/CNFs self-standing electrodes, both viable and flexible. In the first method [46–48], a solution containing active material precursors and a polymer is electrospun and thermal-treated at temperatures suitable to carbonize nanofibers and synthesize the active material. Via the second route [50,51], the solution containing active material precursors is dip-/drop-coated on CNFs obtained by electrospinning, and the fibers are thermal-treated at the synthesis temperature of the active material. In both approaches the active material is synthesized in situ, but some limits can be envisaged: (i) the temperature/atmosphere of the active materials synthesis and nanofibers carbonization mismatching and (ii) the non-homogeneous distribution of the active material along the fibers' thickness by dip/drop-coating.

In this paper, we propose a different approach to synthesize $Na_3MnTi(PO_4)_3$/CNF composites by electrospinning. The active material, synthesized ex situ by the sol–gel route, is added in proper amounts (10 and 30 wt%) to a polyacrylonitrile in N, N-dimethylacetamide solution; the dispersion is electrospun, and the sheets are stabilized and carbonized to obtain self-standing electrodes. The composites are characterized by different techniques (X-ray powder diffraction and Rietveld structural and profile refinement, scanning electron microscopy, energy dispersive spectroscopy, transmission electron microscopy, Raman spectroscopy, and thermogravimetry) to investigate the structure and morphology of and active material distribution in CNFs. The electrochemical performances of the $Na_3MnTi(PO_4)_3$/CNF electrodes are tested and discussed, compared to a $Na_3MnTi(PO_4)_3$ tape-casted electrode (70 wt% active material). An improved capacity at high C-rates and cell lifespan are achieved, thanks to the benefits of CNFs: good electronic conductivity, easy electrolyte permeation into the self-standing electrode, and porous fibers' ability to buffer the active material volume change during sodiation/desodiation.

2. Materials and Methods

2.1. Materials

Sodium acetate (CH_3COONa; Aldrich, Milan, Italy, 99%), manganese (II) acetate tetrahydrate ($(CH_3COO)_2Mn{\cdot}4H_2O$; Aldrich, Milan, Italy, 99%), ammonium phosphate monobasic ($NH_4H_2PO_4$; Aldrich, Milan, Italy, 99%), titanium (IV) isopropoxide ($C_{12}H_{28}O_4Ti$; Aldrich, Milan, Italy, 99%), absolute ethanol (C_2H_5OH, Aldrich, Milan, Italy), citric acid ($C_6H_8O_7$; Aldrich, Milan, Italy, 99%), polyacrylonitrile (PAN: $(C_3H_3N)_n$; Aldrich, Milan, Italy, particle size 50 µm, 99.5% AN/0.5% MA), N,N-dimethylacetamide (DMAc: $CH_3CON(CH_3)_2$; Aldrich, Milan, Italy, 99%), Super P carbon, polyvinylidene fluoride (PVdF Kynar), N-methyl-2-pyrrolidone (NMP: Aldrich, Milan, Italy, 99%), 1M sodium perchlorate ($NaClO_4$) in propylene carbonate (PC) 1:1 v:v (Aldrich, Milan, Italy, 98%), and 4-Fluoro-1,3-diocolan-2-one (FEC; Aldrich, Milan, Italy, 99.5%) were employed to synthesize the active material and the self-standing electrodes and to prepare the slurry of the tape-casted electrode and the electrolyte.

2.2. Synthesis of Electrode Materials

The $Na_3MnTi(PO_4)_3$ active material is synthesized via sol–gel [53]. An aqueous solution of CH_3COONa, $(CH_3COO)_2Mn{\cdot}4H_2O$, $NH_4H_2PO_4$, and citric acid is added dropwise to a solution of titanium (IV) isopropoxide in absolute ethanol. The reagents are taken in a stoichiometric amount, and citric acid equals the transition metal moles. The solution is stirred at 80 °C until a gel forms. The gel is dried at 100 °C, ground in an agate mortar, and heat-treated at 650 °C under nitrogen atmosphere for 12 h. From now on, the powder sample is indicated by the code MnTi.

The self-standing electrodes are synthesized by electrospinning. The electrospun dispersion is prepared by adding MnTi active material (10 and 30 wt%) to a solution of PAN (8 wt%) in DMAc [49,51]. Hereafter, we report the detailed synthesis of the two self-standing electrodes (codes: 10%MnTi/CNF and 30%MnTi/CNF). The MnTi powder is ball-milled at 100 rpm for two cycles (20 min each), then 10 wt% of MnTi (0.376 g) or 30 wt% of MnTi (1.128 g) is added to DMAc (50 mL). The suspension is sonicated for 1 h, after which 3.760 g PAN is added, and the suspension is stirred overnight at 60 °C. The solution is electrospun by using the EF050 Starter Kit Electrospinning system of SKE Research Equipment (C/O Leonardino S.r.l, Bollate, MI, Italy), setting the following conditions: 10.5 mL dispersion, 3.5 mL/h flow, 16 gauge needle, applied voltage 16 kV, needle–collector distance 18 cm, and deposition time 3 h. The setting parameters had been previously optimized. Finally, a homemade humidity sensor-included box is built for humidity control: a value lower than 20% is detected during all depositions.

The same electrospinning procedure is applied to a solution of PAN (8 wt%) in DMAc to prepare a pure CNF sample: it is used for comparison with the self-standing electrode's characterization.

The electrospun sheets are removed from the support (aluminum foil) and stabilized in air for 30 min at 100 °C, 30 min at 200 °C, and finally 2 h at 260 °C (heating ramp: 5 °C min^{-1}). They are further heat-treated at 750 °C for 2 h (heating ramp: 10 °C min^{-1}) in nitrogen atmosphere for the carbonization process.

2.3. Materials Characterization

X-ray powder diffraction (XRPD) measurements are performed by using a Bruker D5005 diffractometer with the Cu Kα radiation (40 kV, 40 mA) and a scintillation detector. The patterns are collected in the 18–80° 2θ range with a step size of 0.03° and 22 s/step counting time. The Rietveld structural refinement is applied to the diffraction data. The NASICON-type structure model (S.G. $R\bar{3}c$) is used to determine the main structural parameters. The TOPAS V3.0 software is used [54].

SEM micrographs are collected by a Zeiss EVO MAH10 (Carl Zeiss, Oberkochen, Germany) scanning electron microscope on Au-sputtered samples (20 kV, secondary electron images, working distance 8.5 mm). The microscope is equipped with an energy dispersive detector (X-max 50 mm^2, Oxford Instruments, Oxford, UK) for the EDS analysis.

TEM images are collected on JEOL JEM-1200EXIII equipped with a TEM CCD camera Mega View III transmission electron microscope to highlight the presence of MnTi powder in the CNFs.

The TGA data collection is performed with a TA Q5000 instrument in air in the 20–725 °C temperature range (heating rate: 10 Kmin^{-1}). The technique is used to determine the effective weight percentage of MnTi powder in CNFs.

The Raman measurements are performed employing a microRaman spectrometer, XploRA Plus HORIBA Scientific (Kyoto, Japan), equipped with an Olympus BX43 microscope. Laser red light at 638 nm (90 mW) is used as the excitation source. The incident laser power is tuned by a set of neutral filters with different optical densities. The investigated samples are placed on a motorized xy stage. The spectral resolution is about 2 cm^{-1}. An open electrode CCD camera, with a multistage Peltier air-cooling system, is used as a detector. The measurements are performed using a 50× objective with a long working distance, which leads to a spatial resolution of the order of 4 μm. The spectra have been collected with a mean integration time of about 10 s and a number of accumulations equal to 10. All the reported data are the result of the average of different spectra collected at different points in each sample.

2.4. Electrochemical Characterization

A Swagelok cell is used for the electrochemical investigation. The cells are assembled in an argon-filled dry box (M. Braun $H_2O < 0.1$ ppm; $O_2 < 0.1$ ppm) by using the self-standing electrodes (see Section 2.2), 1 M $NaClO_4$ in PC and 5% FEC as the electrolyte, and sodium foil as the counter-electrode.

For comparison, a Swagelok cell is assembled by using a tape-casted electrode: to prepare the $Na_3MnTi(PO_4)_3$ slurry, the MnTi powder is ball-milled at 100 rpm for two cycles (20 min each). A mixture of 70 wt% active material, 20 wt% Super P carbon and 10 wt% PVdF is stirred in NMP for 2 h, cast on aluminum foil, and dried at 70 °C for 3 h.

The electrochemical properties are investigated at ambient temperature by cyclic voltammetry (CV) and galvanostatic charge/discharge cycles. The CV is performed with an Autolab potentiostat. All cells are cycled in the 1.5–4.5 V potential range. Galvanostatic charge/discharge cycles are obtained with a Neware-4000BTS apparatus at different current rates in the abovementioned potential range. The electrochemical impedance spectroscopy (EIS) measurements were performed on an Autolab PGSTAT30 potentiostat (Eco Chemie). The EIS spectra were acquired at OCV in the 10^5–10^{-2} Hz frequency range with an amplitude potential of 1 mV.

3. Results and Discussion

3.1. Structural Characterization

In Figure 1, the diffraction patterns of all samples are shown. The MnTi sample displays the diffraction peaks of the NASICON-type crystal structure and well compares to the literature data [53]; no impurity phases are detected. The pure CNF sample displays a broad band at about 25°/2θ, typical of amorphous components. In the MnTi/CNF samples, the peaks of the NASICON structure and the amorphous CNF phase are detected.

Rietveld refinement is applied to the MnTi, 10%MnTi/CNF and 30%MnTi/CNF diffraction patterns. The structural model reported by Zhou and co-workers [53] is used ($R\overline{3}c$ S.G. and lattice parameters a = 8.73352 Å and c = 21.84703 Å). The lattice parameters, crystallite size, and degree of crystallinity obtained by Rietveld refinement are reported in Table S1 (Supporting Information). Figure S1 (Supporting Information) compares the experimental and calculated patterns.

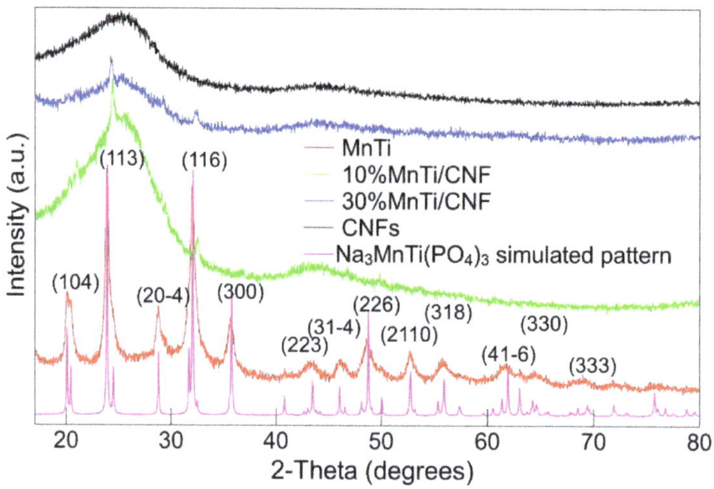

Figure 1. X-ray diffraction patterns of the MnTi, MnTi/CNF, and pure CNF samples. The $Na_3MnTi(PO_4)_3$ simulated pattern is also shown.

The discrepancy factor values (Table S1) and the graphical comparison (Figure S1) demonstrate that the refined model properly fits the experimental XRPD data. The refined lattice parameters of the MnTi sample well compare to the literature ones [53]. However, for both MnTi/CNF samples, the *a* lattice parameter and cell volume decrease, giving rise to a *c/a* ratio increase (Table S1). The cell volume shrinkage may depend on the sintering temperature of 750 °C used for the carbonization process and not on the $Na_3MnTi(PO_4)_3$ inclusion in carbon nanofibers. In fact, the same shrinkage was detected for $Na_3MnTi(PO_4)_3$ samples synthesized at the same temperature by Liu and coworkers [55]. Moreover, the temperature higher than that usually chosen in the sol–gel synthesis (650 °C) may increase intrinsic anti-sites defects, responsible for the charge/discharge voltage hysteresis [39]. The cell shrinkage inhibits the Na(2) reversible insertion/extraction and impacts the electrochemical performances, as discussed in 3.5. However, we chose the carbonization temperature of 750 °C based on the increased conductivity of pure CNFs: 1.56×10^{-3} S/cm and 1.76×10^{-6} S/cm for carbonization at 750 °C and 650 °C, respectively. The $Na_3MnTi(PO_4)_3$ active material in both MnTi and MnTi/CNF samples is nanocrystalline (Table S1), and the crystallite size does not depend on the sintering temperature. The degree of crystallinity gives an idea of the amount of crystalline phase (active material) loaded in the self-standing electrodes. The degree of crystallinity is about 9% for the MnTi/CNF samples (Table S1). The value matches the synthesis content of the 10%MnTi/CNF but not the 30%MnTi/CNF one. It may depend on the possible decantation of MnTi powder into the tube during the electrospinning deposition. The quantitative amount of the active material loaded into CNFs will be evaluated by TGA and discussed in 3.3.

3.2. Samples Morphology

Figure 2a,b shows the SEM images of the MnTi sample. Large aggregates (10–30 µm), composed of sub-particles smaller than 1 µm, are observed. The grain's surfaces are rough but well defined. The TEM micrographs in Figure 2c,d confirm the presence of the particles' aggregates. Noteworthy, the particulate is surrounded by homogeneously spread carbon coating due to the carbon source (citric acid) used in the sol–gel synthesis.

Figure 2. SEM (**a**,**b**) and TEM (**c**,**d**) images of the MnTi sample.

The SEM surface and cross-section images of the 10%MnTi/CNF sheets (Figure 3a–c) demonstrate that the $Na_3MnTi(PO_4)_3$ is dispersed into nanofibers and forms agglomerates with a widely varying size distribution. The 10%MnTi/CNF sheet thickness is about 50 µm. As in the case of the 10%MnTi/CNF, the SEM surface and cross-section images of the 30%MnTi/CNF sample (Figure 3d–f) display agglomerates with an uneven size spread in the CNF matrix. The sheet thickness is about 300 µm. The carbonization process does not influence the homogeneous distribution of the active material, as demonstrated by the SEM images taken on the electrospun and graphitized 10%MnTi/CNF sample (Figure S2). As expected, the carbonization process causes a slight decrease in CNFs' diameter.

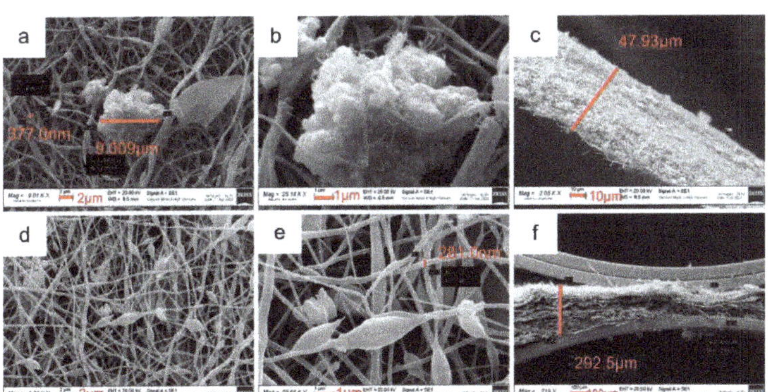

Figure 3. SEM images of surface (**a**,**b**) and cross-section (**c**) of the 10%MnTi/CNF sample. SEM images of surface (**d**,**e**) and cross-section (**f**) of the 30%MnTi/CNF sample.

The morphology, particle size, and distribution of the active material in self-standing electrodes are deeply investigated by TEM analysis. The TEM images of the 10%MnTi/CNF sample (Figure 4a–c) display $Na_3MnTi(PO_4)_3$ nanoparticles of about 20–30 nm, consistent with the crystallite size reported in Table S1; they form aggregates of variable size and segregate between and into nanofibers (CNFs diameter: 120–170 nm). The 30%MnTi/CNF sample displays comparable morphology (Figure 4d–f), and the $Na_3MnTi(PO_4)_3$ aggregates are spread both between and embedded into nanofibers. The CNFs' diameter ranges between 90 and 120 nm. In the MnTi/CNF samples, the agglomerates display widely varying sizes, as also suggested by the granulometric study based on the SEM data (mean particle size: 0.57(0.54) μm).

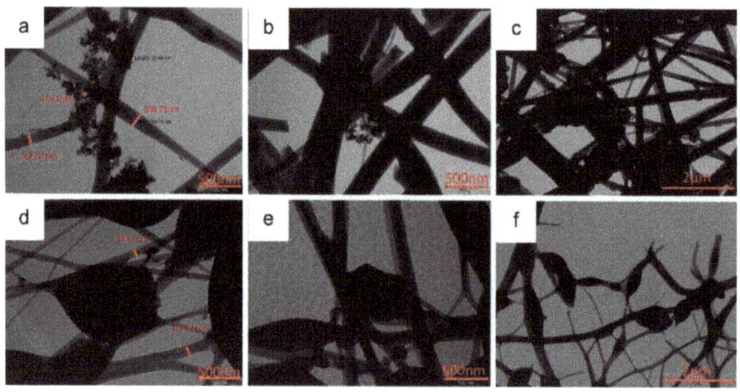

Figure 4. TEM images of 10%MnTi/CNF (**a–c**) and 30%MnTi/CNF (**d–f**) samples taken at different magnifications.

The surface and bulk distribution of MnTi powder in the MnTi/CNF self-standing electrodes is investigated by EDS. The Na, Mn, Ti, and P distribution maps on the surface of the 10%MnTi/CNF sample (Figure 5a–e) confirm that the active material aggregates within and between CNFs and that it is homogeneously spread in CNFs. What is noteworthy is that the particles of the active material are spread along the sheet thickness, as demonstrated by the cross-section element distribution maps (Figure 5f–l); this is beneficial to obtain enhanced electrochemical performances, and it is not easily achieved by the dip- and drop-coating loading approach. Comparable EDS results are obtained for 30%MnTi/CNF, as demonstrated by the element distribution maps on the surface (Figure 6a–e) and cross-section (Figure 6f–l).

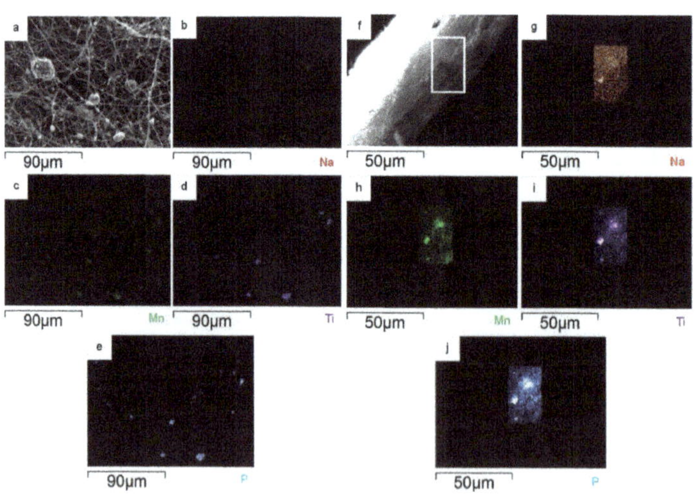

Figure 5. On the left surface: portion of 10%MnTi/CNF sample (**a**) and distribution maps of Na (**b**), Mn (**c**), Ti (**d**), and P (**e**). On the right: cross-section portion of 10%MnTi/CNF sample (**f**) and distribution maps of Na (**g**), Mn (**h**), Ti (**i**), and P (**j**).

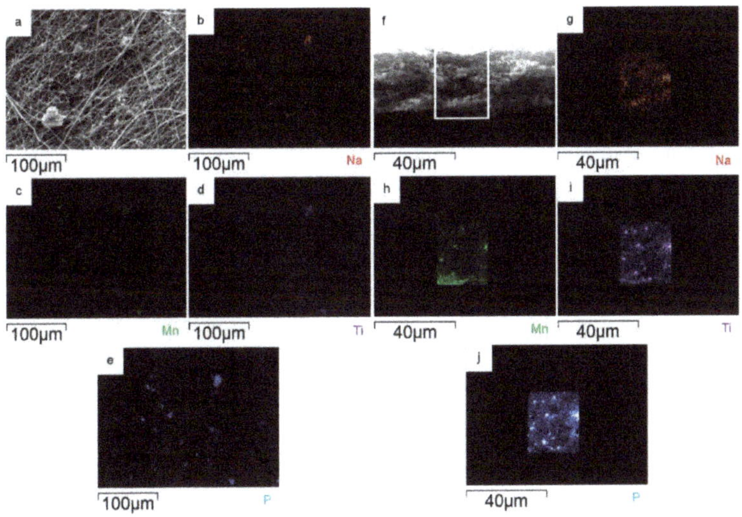

Figure 6. On the left: surface portion of 30%MnTi/CNF sample (**a**) and distribution maps of Na (**b**), Mn (**c**), Ti (**d**), and P (**e**). On the right: cross-section portion of 30%MnTi/CNF sample (**f**) and distribution maps of Na (**g**), Mn (**h**), Ti (**i**), and P (**j**).

3.3. Thermogravimetric Analysis

The carbon content and the $Na_3MnTi(PO_4)_3$ amount loaded into CNFs is evaluated by thermogravimetric analysis on the samples after the carbonization process. The thermogravimetric curves of MnTi, 10%MnTi/CNF, and 30%MnTi/CNF samples are shown in Figure 7.

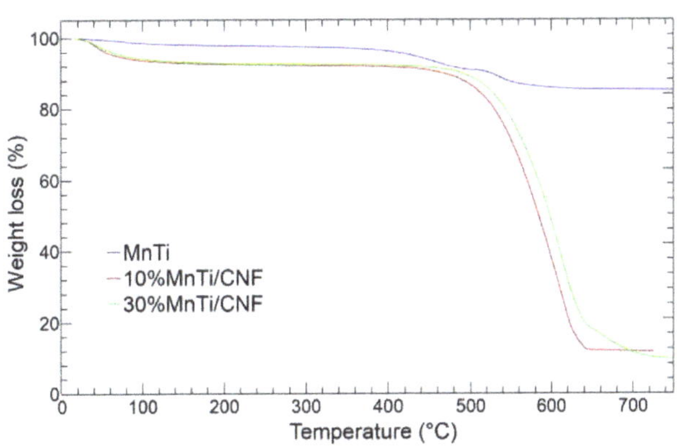

Figure 7. TGA curves of MnTi (blue), 10%MnTi/CNF (red), and 30%MnTi/CNF (green) samples. The analysis is performed in air between 25 and 730 °C.

The MnTi TGA curve displays three weight losses: 1.43 wt% at 100 °C, 13.88 wt% at 400–500 °C, and 0.89 wt% at 600 °C. The first loss is due to the release of adsorbed water, and the second and third account for the carbon coating content in the MnTi sample: 14.77 wt%. Finally, the small weight increase at 500 °C is explained by the oxidation of the low-valence state of metal species induced by carbon combustion, as reported in the literature [40,53].

The TGA curves of the 10%MnTi/CNF (red) and 30%MnTi/CNF (green) samples also show two mass losses: (i) below 100 °C due to the adsorbed water release (about

7.7 wt% for both samples); (ii) in the 500–650 °C temperature range, involving the carbonaceous component combustion of both the $Na_3MnTi(PO_4)_3$ carbon coating and the carbon nanofibers. The latter mass loss corresponds to the carbon content in the samples: 80.57 wt% for 10%MnTi/CNF and 82.71 wt% for 30%MnTi/CNF. The 10%MnTi/CNF and 30%MnTi/CNF samples give a residual mass of 11.73 wt% at 650 °C and 9.59 wt% at 700 °C, respectively, which corresponds to the active material amount in the samples. The result matches the synthesis value of the 10%MnTi/CNF self-standing electrode. In the case of the 30%MnTi/CNF sample, the $Na_3MnTi(PO_4)_3$ amount is much lower than the synthesis value (30 wt%). The TGA results match the values of the degree of crystallinity obtained by XRPD analysis (Table S1). We repeated the synthesis of the 30%MnTi/CNF sample, and we noticed that some sedimentation of the active material occurred along the tube connecting the pump and the needle of the horizontal spinneret. We believed this may have been due to the possible poor dispersibility of the pre-synthesized active material aggregates in the PAN-based polymer solution. We also tried to prepare the self-standing electrode using vertical equipment, but again, the $Na_3MnTi(PO_4)_3$ 30 wt% amount was not achieved. Based on the above-mentioned results, we decided to investigate the electrochemical performances of the MnTi (tape-casted) and 10%MnTi/CNF (self-standing) electrodes.

3.4. Raman Analysis

The Raman analysis is performed to obtain insights on the structural compositions of the investigated materials. In particular, the aim is to study the Raman activity of carbon modes in both carbon coatings and CNFs, thus determining the order degree of the carbon component itself.

The room temperature spectra of the 10%MnTi/CNF and 30%MnTi/CNF self-standing samples together with the ones of the MnTi powder and the only CNFs are reported in Figure 8.

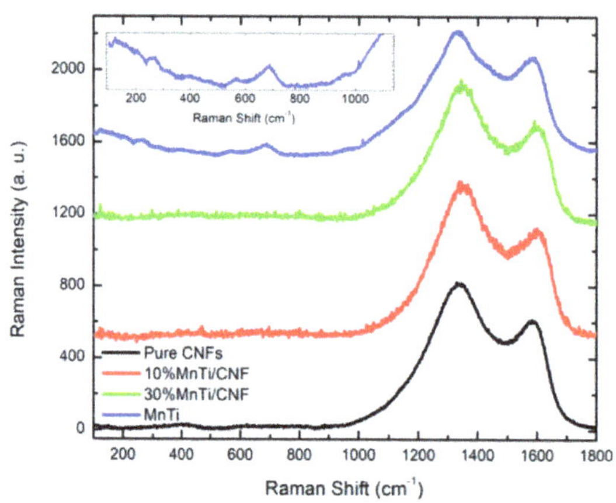

Figure 8. Room temperature Raman spectra of pure CNFs, 10%MnTi/CNF, 30%MnTi/CNF, and MnTi powder (from bottom to top).

The MnTi spectrum displays some broad signals in the low-frequency region (see inset). According to [56], the most prominent features at about 565 and 680 cm^{-1} could be ascribed to P–O and P–O–Na vibrations together with possible contribution by Ti–O ones, while less intense modes are visible below 300 cm^{-1} and they could arise from the vibrations involving Mn ions.

Apart from the MnTi sample that shows low-energy modes, each spectrum is dominated only by two intense modes that fall at about 1340 and 1585 cm^{-1}. These features are a characteristic signature of carbonaceous materials, through which it is possible to obtain information about the order/disorder as well as the crystalline quality of the structure.

Indeed, as it is well known, the mode occurring at about 1585 cm^{-1} is the G band and it is characteristic of an ordered graphitic network, while the one at ≈1340 cm^{-1} is known as the D band and it is generally ascribed to defectiveness [57].

The presence of the D and G bands also in the MnTi spectrum confirms the presence of a carbon coating on the MnTi particles, as evidenced by the morphological analysis and TGA data. The D mode seems to be broader than in the other samples, pointing towards a more disordered structure, consistent with an external citric carbon coating.

In Figure 9, the value of the I_G/I_D ratio is reported for each sample. In general, the I_G and I_D values correspond to the intensities of the modes. However, to better account for the broadening of the peak, especially in the case of the MnTi sample, we have used the integrated intensity of the two Gaussian functions used to perform the best fitting procedure.

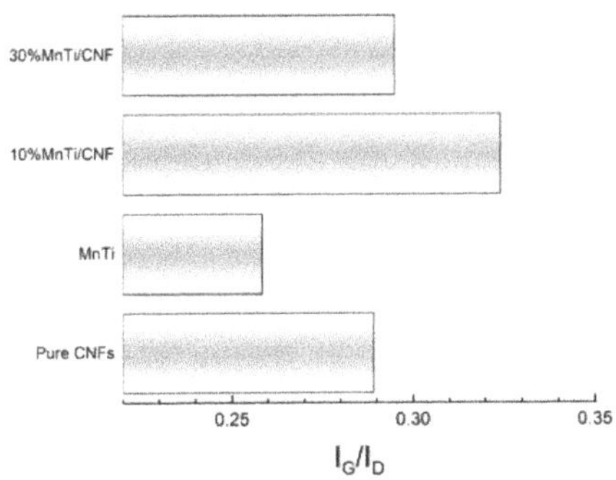

Figure 9. I_G/I_D values reported for each investigated sample.

The smallest I_G/I_D value (0.26) is obtained for the MnTi sample as expected: CNFs are the main source of ordered carbon, beneficial for the electronic conductivity enhancement of the electrodes. What is noteworthy is that a higher I_G/I_D value is detected in the 10%MnTi/CNF sample, which will be electrochemically evaluated. This fact could be due to a beneficial effect of the native carbon coating of MnTi particles which favors an ordered carbon nanofiber adhesion. This effect should be more evident for low amounts of MnTi aggregates due to the weight of different Raman yields in the whole Raman spectrum of the sampled volumes.

3.5. Electrochemical Characterization

The $Na_3MnTi(PO_4)_3$ shows three redox peaks of Ti^{3+}/Ti^{4+}, Mn^{2+}/Mn^{3+}, and Mn^{3+}/Mn^{4+} at 2.1 V, 3.5 V, and 4.0 V, respectively. The presence of three different redox reactions makes the material suitable for different applications that require different applied voltages: as an anode for the Ti^{3+}/Ti^{4+} redox couple and as a cathode thanks to Mn^{2+}/Mn^{3+} and Mn^{3+}/Mn^{4+} ones [53].

The redox mechanism involves a multielectron process during the Na^+ extraction/insertion. It includes a two-electron transfer for Mn^{2+}/Mn^{3+} and Mn^{3+}/Mn^{4+} and one electron process in the case of Ti^{3+}/Ti^{4+}; an ex situ X-ray diffraction investigation demonstrates the sodiation/desodiation process involves both solid–solution and two-phase reactions [41,53]. Equations (1) and (2) summarize the charge/discharge process [41,53]:

$$\text{discharge}: Na_1MnTi(PO_4)_3 \underset{4.0\,V}{\rightarrow} Na_2MnTi(PO_4)_3 \underset{3.5\,V}{\rightarrow} Na_3MnTi(PO_4)_3 \underset{2.1\,V}{\rightarrow} Na_4MnTi(PO_4)_3 \quad (1)$$

$$\text{charge}: Na_4MnTi(PO_4)_3 \underset{2.1\,V}{\rightarrow} Na_3MnTi(PO_4)_3 \underset{3.5\,V}{\rightarrow} Na_2MnTi(PO_4)_3 \underset{4.0\,V}{\rightarrow} Na_1MnTi(PO_4)_3 \quad (2)$$

The $Na_2MnTi(PO_4)_3/Na_1MnTi(PO_4)_3$ process is kinetically favored with the fastest Na^+ diffusion, as also supported by density functional theory calculations [53].

3.5.1. Cyclic Voltammetry

The MnTi CV curve is shown in Figure 10a. The tape-casted electrode displays the three redox peaks of Ti^{3+}/Ti^{4+}, Mn^{2+}/Mn^{3+}, and Mn^{3+}/Mn^{4+} at 2.20 V/2.06 V, 3.67 V/3.41 V, and 4.08 V/3.95 V, respectively. The ΔV of Ti^{3+}/Ti^{4+} peak is 139 mV and progressively diminishes for Mn^{2+}/Mn^{3+} (26 mV) and Mn^{3+}/Mn^{4+} (13 mV). It indicates a quite small polarization phenomenon. The current intensity is higher than 0.04 A/g and lower than −0.06 A/g for anodic and cathodic peaks, respectively. The three redox couples are also consistent with the three plateaus detected in the Galvanostatic charge–discharge profiles shown in Figure 10b. In the first charge process, the plateau of the Ti^{3+}/Ti^{4+} redox couple is not observed due to the initial open-circuit voltage (2.5–2.7 V). The CV curves and charge/discharge profiles of the 10%MnTi/CNF sample are reported in Figure 10c and Figure 10d, respectively. In this case, only the Ti^{3+}/Ti^{4+} redox peaks at 2.12 V/2.06 V are detected, while the Mn^{2+}/Mn^{3+} and Mn^{3+}/Mn^{4+} ones become very faint. The sluggish redox activity may be attributed to the cell shrinkage at a sintering temperature of 750 °C [55] and to the intrinsic anti-site defect: Mn occupies the Na(2) vacancy and hampers the sodium ion diffusion and the manganese redox processes [39]. However, a temperature of 750 °C is necessary to obtain a successful carbonization. Secondly, we underline that the 10%MnTi/CNF electrode contains only 10 wt% of active material against the 70 wt% present in the MnTi tape-casted electrode, and this can influence the CV. The Ti^{3+}/Ti^{4+} redox reaction is also confirmed by the evident plateau in Figure 10d. In the CV curves (Figure 10c), the weak peak at 2.4 V is explained by the disproportion reaction of Mn^{3+} dissolved in the electrolyte [41]. Finally, the cycles are overlapped to demonstrate a strong redox reversibility.

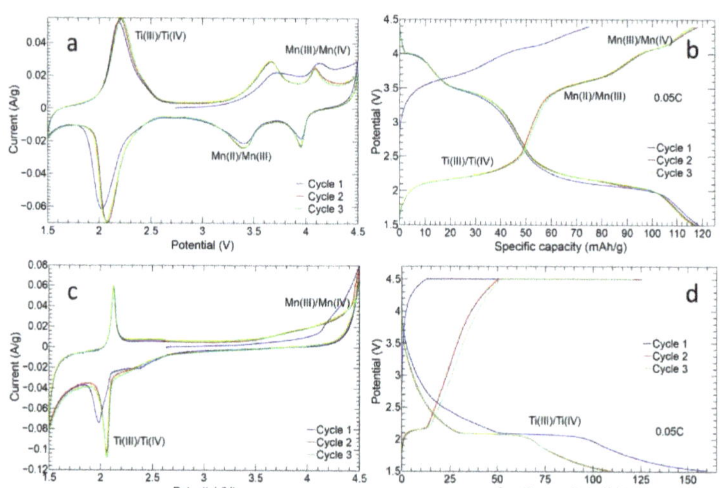

Figure 10. Cyclic voltammetry curves and first three charge/discharge profiles of MnTi tape-casted (**a**,**b**) and 10%MnTi/CNF (**c**,**d**) electrodes. The analysis is performed at 0.1 mV/s in the 1.5–4.5 V range.

Figure 11a,d show the CV curves at different scan rates for MnTi and 10%MnTi/CNF samples, respectively. The data were analyzed to evaluate the sodium ion diffusion coefficient, the alkali metal-ion faradaic contribution (diffusion control), and the non-faradaic one (pseudo-capacitance control) caused by the formation of the double layer at the material surface [41,53,55]. The pseudo-capacitance contributions are shown in Figure 11b,c,e,f for the MnTi and 10%MnTi/CNF samples, respectively. The relationship between the redox current i_p (A) and scan rate ν (mV/s) is given by the following:

$$i_p = k_1\nu + k_2\sqrt{\nu} \tag{3}$$

where i_p is the peak current, ν is the scan rate, and k_1 and k_2 are adjustable parameters [41,53]. The diffusion contribution is driven by the square root term thanks to the derivation of the i_p by the Randles–Sevick equation:

$$i_p = 2.69 \times 10^5 n^{\frac{3}{2}} AC\sqrt{D\nu} \tag{4}$$

where n is the number of electrons transferred, A is the electrode area, C is the Na^+ concentration and D is the diffusion coefficient. The quantities are given in the CGS unit system and at standard conditions. From Equations (3) and (4), the diffusion coefficient D can be evaluated by calculating the k_2 term for both MnTi and 10%MnTi/CNF samples.

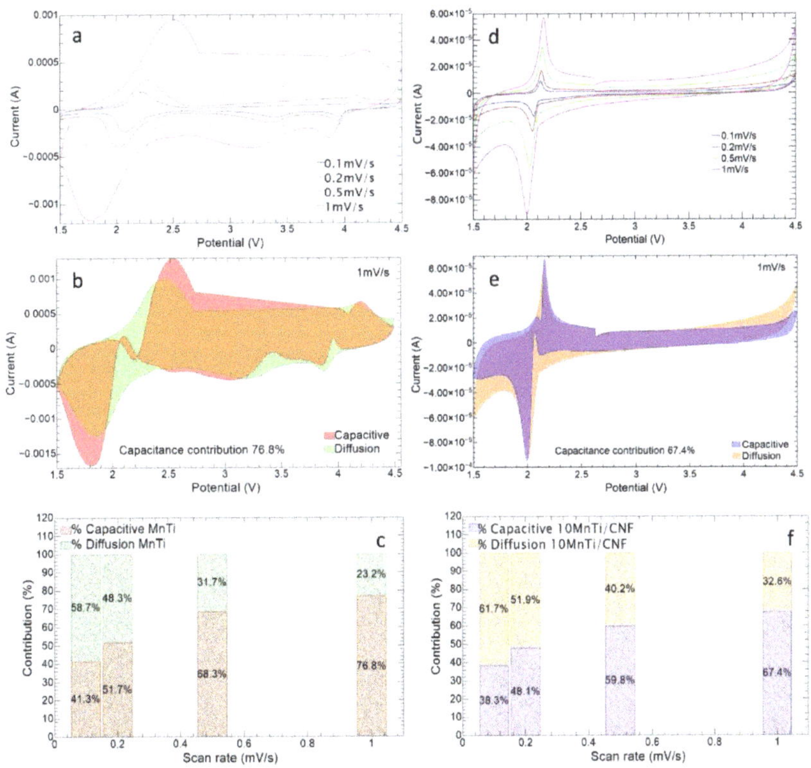

Figure 11. Cyclic voltammetry at different scan rates for MnTi tape-casted (**a**) and 10%MnTi/CNF (**d**) electrodes. Capacitive and diffusive contributions at 1 mV/s for MnTi tape-casted (capacitive: red; diffusive: green; both contributions: light brown) (**b**) and 10%MnTi/CNF (capacitive: blue; diffusive: orange; both contributions: violet) (**e**) electrodes. Capacitance and diffusion histogram for MnTi tape-casted (**c**) and 10%MnTi/CNF (**f**).

The tape-casted electrode (A = 0.78 cm^2) exhibits diffusion coefficients values for anodic peaks of $D_{Ti(III)/Ti(IV)} = 1.3 \times 10^{-9}$ cm^2/s, $D_{Mn(II)/Mn(III)} = 8.4 \times 10^{-9}$ cm^2/s, and $D_{Mn(III)/Mn(IV)} = 1.6 \times 10^{-10}$ cm^2/s. In the case of the cathodic peak, the D values are $D_{Ti(III)/Ti(IV)} = 2.2 \times 10^{-9}$ cm^2/s, $D_{Mn(II)/Mn(III)} = 7.6 \times 10^{-9}$ cm^2/s, and $D_{Mn(III)/Mn(IV)} = 1.4 \times 10^{-9}$ cm^2/s. The results compare to the literature ones [53,55].

An equivalent analysis can be conducted for the self-standing electrode as well. Contrary to the MnTi, the 10%MnTi/CNF electrode features a three-dimensional structure, whose main advantage is the substantial increase in the active material surface. Indeed, 10%MnTi/CNF area A cannot be estimated from the electrode section. Therefore, we evaluate an equivalent anodic diffusion coefficient $D_{eq} = 2.6 \times 10^{-12}$ cm^2/s and cathodic $D_{eq} = 1.0 \times 10^{-11}$ cm^2/s for Ti^{3+}/Ti^{4+} redox couple, where we assume electrode area A is equal to its section. The result is consistent with the fact that the Ti^{3+}/Ti^{4+} self-standing electrode shows a lower current intensity than its tape-casted counterpart (see Figure 11a,d), and this depends on the $Na_3MnTi(PO_4)_3$ amount in the 10%MnTi/CNF and MnTi electrodes (10 and 70 wt%, respectively). The low amount of the active material leads to a low electrochemical active area.

To compare the performance of tape-casted and self-standing electrodes, the capacity contribution of diffusion and reaction (capacitive contribution) control as a function of CV sweep rates were calculated. As shown in Figure 11e,f, the 10%MnTi/CNF sample exhibits a higher contribution of diffusion control at each scan rate than its tape-casted counterpart (Figure 11b,c), thanks to the presence of the very porous CNF sheets. Indeed, the porosity of non-woven nanofibers guarantees an easier electrolyte diffusion which easily makes contact with active material particles, as reported in the literature [58,59]. This is also confirmed by the complete electrolyte permeation of the 10%MnTi/CNF sheet after carbonization, as shown in Figure S3 (Supporting Information). Finally, the tape-casted anodic and cathodic peaks tend to move to the right and left, respectively as the scan rate increases (Figure 11a), while this behavior is less pronounced for the self-standing sample (Figure 11d). For example, the Ti^{3+}/Ti^{4+} redox peak overpotentials of the MnTi sample are 139 mV, 280 mV, 503 mV, and 768 mV as the scan rate increases against 63 mV, 58 mV, 119 mV, and 154 mV for the 10%MnTi/CNF. The results suggest more irreversible redox processes in the tape-casted electrode than in the self-standing one.

3.5.2. Charge/Discharge Cycles

Figure 12 shows the charge/discharge analysis for both MnTi and 10%MnTi/CNF samples.

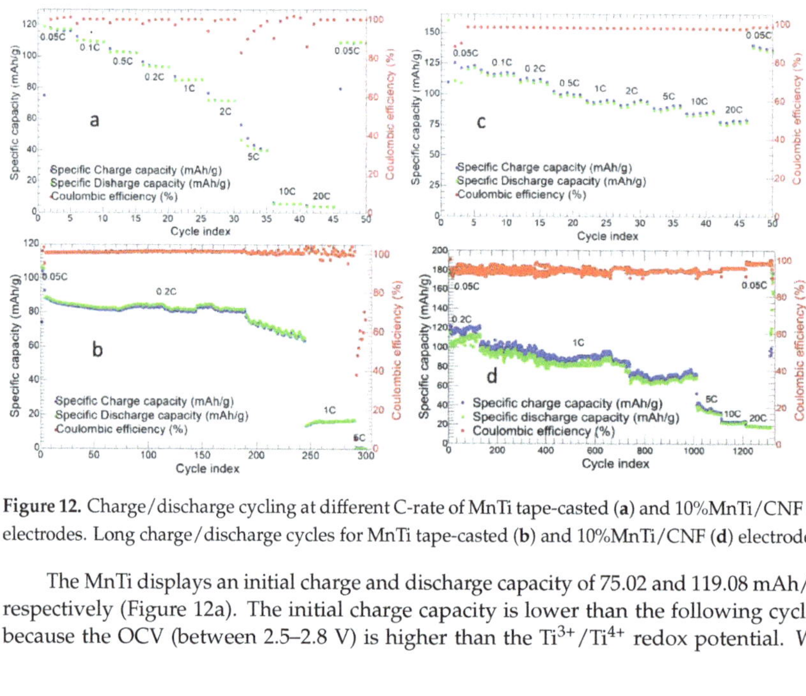

Figure 12. Charge/discharge cycling at different C-rate of MnTi tape-casted (**a**) and 10%MnTi/CNF (**c**) electrodes. Long charge/discharge cycles for MnTi tape-casted (**b**) and 10%MnTi/CNF (**d**) electrodes.

The MnTi displays an initial charge and discharge capacity of 75.02 and 119.08 mAh/g, respectively (Figure 12a). The initial charge capacity is lower than the following cycles because the OCV (between 2.5–2.8 V) is higher than the Ti^{3+}/Ti^{4+} redox potential. We obtain average discharge capacities of 117.76, 109.80, 102.69, 93.67, 85.09, 71.83, 42.03, 5.70, and 4.12 mAh/g at 0.05C, 0.1C, 0.2C, 0.5C, 1C, 2C, 5C, 10C, and 20C, respectively. The cell exhibits a good capacity recovery at the end of the measurement and a coulombic efficiency ≥98%. In Figure 12b, the long charge/discharge cycles are shown. After the first five cycles at 0.05C, the cell is tested at 0.2C, 1C, and 5C for 240, 50, and 20 cycles, respectively. The initial discharge capacity is 106.48 mAh/g, and the cell exhibits a coulombic efficiency ≥99%. The specific capacity decreases, increasing both the cycle index and C-rate. For the first 190 cycles at 0.2C, the average capacity value is 84.24 mAh/g, while it diminishes to 71.63 mAh/g in the following 50 ones at the same C-rate with a capacity retention of 73.97% at the 240th cycle. By increasing the C-rate from 0.2C to 1C, the capacity value decreases to 16.18 mAh/g, while at 5C, the cell does not work. So, it can be concluded that the MnTi sample does not support C-rates higher than 1C after 290 cycles.

The 10%MnTi/CNF shows an initial charge and discharge capacity of 109.9 and 160.04 mAh/g, respectively (see Figure 12c). We obtain average discharge capacities of 124.38, 115.68, 111.04, 100.68, 93.6, 91.42, 89.15, 88.30, and 77.60 mAh/g at 0.05C, 0.1C, 0.2C, 0.5C, 1C, 2C, 5C, 10C, and 20C, respectively. At the end of the measurement, the cell completely recovers the initial capacity with a coulombic efficiency ≥98%. Contrary to the tape-casted MnTi electrode, the self-standing 10%MnTi/CNF one exhibits (i) a very moderate capacity loss by increasing the C-rate, (ii) good stability and reversibility of sodium storage, and (iii) promising performances also at C-rates higher than 5C. The 10%MnTi/CNF ability to cycle at high C-rates is guaranteed by the CNFs 3D network that gives high porosity and an easier diffusion of Na-ion. This implies a high power density, as reported by Vu and co-workers [58]. The capacity values of the 10%MnTi/CNF electrode shown in Figure 12c are comparable to the literature ones [40,41]. Notably, these promising electrochemical performances are obtained on an electrode synthesized by a simple and feasible approach with an active material amount of 10 wt%, against the 70–80 wt% of the conventional tape-casted electrodes. We underline some drawbacks can be envisaged in the sintering temperature of 750 °C necessary for the CNFs' graphitization as this implies cell shrinkage and the sluggishness of the redox process. Nevertheless, they are overcome thanks to the advantages of the CNFs: they provide conducive carbon, a porous matrix beneficial for electrolyte–active material contact, easy sodium ions diffusion, and a light and self-standing electrode.

The very promising electrochemical performances of the 10%MnTi/CNF sample are also confirmed by the long-term charge/discharge cycling shown in Figure 12d: the cell lifespan is tested at 0,05C, 0.2C, 1C, 5C, 10C, and 20C for 5, 50, 1000, 100, 100, and 100 cycles, respectively. At 0.05C, the discharge capacity of 173.4 mAh/g in the first cycle approaches the theoretical one, and then values of about 120 mAh/g are achieved. The specific capacity decreases, increasing the cycle index and at the C-rate change. The specific capacity ranges between 67.3 mAh/g and 110.1 mAh/g for 1000 cycles at 1C. Notably, after 1000 cycles at 1C, the cell can be further cycled at higher C-rates with a final capacity retention of 59.6%. Contrary to the tape-casted electrode for which the capacity dramatically decreases at 1C (16.18 mAh/g) after 250 cycles at 0.2C (Figure 12b), the self-standing 10%MnTi/CNF electrode shows a longer lifespan, and its capacity is completely recovered in the last ten cycles at 0.05C after 1350 cycles. The good long-term cycling is another advantage of using CNFs for the electrode. As reported by the Vu group [58], the porosity facilitates the diffusion of the ions in bulk electrodes and also buffers the volume change during the charge and discharge cycle, providing a longer cell lifespan. The coulombic efficiency is ≥93% which implies a lower redox reversibility, especially in the first 50 cycles. Upon increasing the C-rate, the specific capacity is 33.6, 22.6, and 17.3 mAh/g at 5C, 10C, and 20C, respectively. Finally, we performed electrochemical impedance spectroscopy measurements on the cycled tape-casted and 10%MnTi/CNF self-standing electrodes. The Nyquist plot is shown in Figure S4. The impedance spectra compare to the literature ones for the $Na_3MnTi(PO_4)_3$ active material with different carbon coatings (with and without graphene oxide) [40]. The smaller diameter of the semicircle in the high-frequency region for the

10%MnTi/CNF electrode confirms smaller charge transfer resistance (585.7 Ω vs. 803.8 Ω of the tape-casted cathode) and a faster charge transfer at the electrode–electrolyte interface. The larger slope of the Warburg impedance of the 10%MnTi/CNF indicates more favorable Na ion transport in the self-standing electrode compared to the tape-casted one.

4. Conclusions

In this study, a simple and viable approach is used to synthesize self-standing electrodes for SIBs based on $Na_3MnTi(PO_4)_3$ active material loaded into carbon nanofibers by electrospinning. The CNFs demonstrate to be a suitable matrix to host $Na_3MnTi(PO_4)_3$ particles. The active material is detected both in and within CNFs and it is homogeneously distributed along the sheet thickness. The loaded $Na_3MnTi(PO_4)_3$ maintains its NASICON-type crystal structure, but the sintering temperature of 750 °C used for the carbonization process induces cell shrinkage. While this implies a sluggish redox activity, the presence of the very porous non-woven nanofibers guarantees an easier electrolyte diffusion and an increased alkali metal-ion faradaic contribution. The charge/discharge cycling tests at different C-rates and long-term cycling investigations confirm the promising electrochemical performances of the self-standing electrode compared to its conventional tape-casted counterpart. The 10%MnTi/CNF electrode displays an initial discharge capacity of 124.38 mAh/g at 0.05C, which is completely recovered at the end of the measurement with a coulombic efficiency ≥98%. The capacity value at 20C is 77.60 mAh/g. The self-standing electrode gives an improved lifespan compared to the tape-casted one: it exhibits capacities in the 67.2–110.1 mAh/g range at 1C and can further be cycled at 5C, 10C, and 20C after 1000 cycles at 1C (total cycles: 1350), contrary to the tape-casted one working only for 300 cycles and up to 5C. Notably, the enhanced capacity and cycling performances are obtained by only 10 wt% of active material loading into CNFs compared to 70 wt% of its tape-casted counterpart.

Supplementary Materials: The following supporting information can be downloaded at: https://www.mdpi.com/article/10.3390/nano14090804/s1, Figure S1: Rietveld refinement of the X-ray diffraction data of the Na3MnTi(PO4)3 pristine material and CNF-loaded Na3MnTi(PO4)3 samples; Figure S2: SEM images of the electrospun and graphitized 10%MnTi/CNF sample; Figure S3: electrolyte permeation in 10%MnTi/CNF sample; Figure S4: Nyquist plot of the 10%MnTi/CNF and MnTi electrodes; Table S1: Lattice parameters, cell volume, *c/a* ratio, crystallite size, crystallinity percentage, and discrepancy factors obtained by the Rietveld refinement of the diffraction data of the pristine and $Na_3MnTi(PO_4)_3$/CNF samples.

Author Contributions: Conceptualization, D.M.C. and D.C.; formal analysis, D.M.C., P.G. and D.C.; investigation, D.M.C., C.U., G.B., B.A., V.B. and A.G.; resources, D.C.; supervision, D.C.; visualization, D.M.C., C.U. and B.A.; writing—original draft preparation, D.M.C., P.G. and D.C.; writing—review and editing, D.M.C., C.U., P.G. and D.C. All authors have read and agreed to the published version of the manuscript.

Funding: This research received no external funding.

Data Availability Statement: The data presented in this study are available on request from the corresponding author.

Acknowledgments: The authors are grateful to Massimo Boiocchi for their support in the TEM analysis performed at the Centro Grandi Strumenti, the University of Pavia, to Daniele Spada of the Department of Chemistry, Physical Chemistry section, the University of Pavia, for their support in the electrochemical data analysis, and to Matteo Lugli of the Department of Physics, the University of Pavia, for the fabrication of a homemade humidity control for the electrospinning process. The authors acknowledge support from the Ministero dell'Università e della Ricerca (MUR) and the University of Pavia through the program "Dipartimenti di Eccellenza 2023–2027".

Conflicts of Interest: The authors declare no conflicts of interest.

References

1. Larcher, D.; Tarascon, J.-M. Towards Greener and More Sustainable Batteries for Electrical Energy Storage. *Nat. Chem.* **2015**, *7*, 19–29. [CrossRef]
2. Kittner, N.; Lill, F.; Kammen, D.M. Energy Storage Deployment and Innovation for the Clean Energy Transition. *Nat. Energy* **2017**, *2*, 17125. [CrossRef]
3. Li, H.; Zhou, X.; Zhai, W.; Lu, S.; Liang, J.; He, Z.; Long, H.; Xiong, T.; Sun, H.; He, Q.; et al. Phase Engineering of Nanomaterials for Clean Energy and Catalytic Applications. *Adv. Energy Mater.* **2020**, *10*, 2002019. [CrossRef]
4. Dunn, B.; Kamath, H.; Tarascon, J.-M. Electrical Energy Storage for the Grid: A Battery of Choices. *Science* **2011**, *334*, 928–935. [CrossRef]
5. Yang, Z.; Zhang, J.; Kintner-Meyer, M.C.W.; Lu, X.; Choi, D.; Lemmon, J.P.; Liu, J. Electrochemical Energy Storage for Green Grid. *Chem. Rev.* **2011**, *111*, 3577–3613. [CrossRef] [PubMed]
6. Goodenough, J.B. Evolution of Strategies for Modern Rechargeable Batteries. *Acc. Chem. Res.* **2013**, *46*, 1053–1061. [CrossRef] [PubMed]
7. Scrosati, B.; Hassoun, J.; Sun, Y.-K. Lithium-Ion Batteries. A Look into the Future. *Energy Environ. Sci.* **2011**, *4*, 3287–3295. [CrossRef]
8. Gao, C.; Dong, Q.; Zhang, G.; Fan, H.; Li, H.; Hong, B.; Lai, Y. Antimony-Doped Lithium Phosphate Artificial Solid Electrolyte Interphase for Dendrite-Free Lithium-Metal Batteries. *ChemElectroChem* **2019**, *6*, 1134–1138. [CrossRef]
9. Huang, Y.; Yang, H.; Xiong, T.; Adekoya, D.; Qiu, W.; Wang, Z.; Zhang, S.; Balogun, M.-S. Adsorption Energy Engineering of Nickel Oxide Hybrid Nanosheets for High Areal Capacity Flexible Lithium-Ion Batteries. *Energy Storage Mater.* **2020**, *25*, 41–51. [CrossRef]
10. Balogun, M.-S.; Zeng, Y.; Qiu, W.; Luo, Y.; Onasanya, A.; Olaniyi, T.K.; Tong, Y. Three-Dimensional Nickel Nitride (Ni_3N) Nanosheets: Free Standing and Flexible Electrodes for Lithium Ion Batteries and Supercapacitors. *J. Mater. Chem. A* **2016**, *4*, 9844–9849. [CrossRef]
11. Balogun, M.-S.; Wu, Z.; Luo, Y.; Qiu, W.; Fan, X.; Long, B.; Huang, M.; Liu, P.; Tong, Y. High Power Density Nitridated Hematite (α-Fe_2O_3) Nanorods as Anode for High-Performance Flexible Lithium Ion Batteries. *J. Power Sources* **2016**, *308*, 7–17. [CrossRef]
12. Whittingham, M.S. Lithium Batteries and Cathode Materials. *Chem. Rev.* **2004**, *104*, 4271–4302. [CrossRef] [PubMed]
13. Goodenough, J.B.; Park, K.-S. The Li-Ion Rechargeable Battery: A Perspective. *J. Am. Chem. Soc.* **2013**, *135*, 1167–1176. [CrossRef] [PubMed]
14. Kundu, D.; Talaie, E.; Duffort, V.; Nazar, L.F. The Emerging Chemistry of Sodium Ion Batteries for Electrochemical Energy Storage. *Angew. Chem. Int. Ed.* **2015**, *54*, 3431–3448. [CrossRef] [PubMed]
15. Delmas, C. Sodium and Sodium-Ion Batteries: 50 Years of Research. *Adv. Energy Mater.* **2018**, *8*, 1703137. [CrossRef]
16. Slater, M.D.; Kim, D.; Lee, E.; Johnson, C.S. Sodium-Ion Batteries. *Adv. Funct. Mater.* **2013**, *23*, 947–958. [CrossRef]
17. Yabuuchi, N.; Kubota, K.; Dahbi, M.; Komaba, S. Research Development on Sodium-Ion Batteries. *Chem. Rev.* **2014**, *114*, 11636–11682. [CrossRef] [PubMed]
18. You, Y.; Wu, X.-L.; Yin, Y.-X.; Guo, Y.-G. High-Quality Prussian Blue Crystals as Superior Cathode Materials for Room-Temperature Sodium-Ion Batteries. *Energy Environ. Sci.* **2014**, *7*, 1643–1647. [CrossRef]
19. Lee, H.-W.; Wang, R.Y.; Pasta, M.; Woo Lee, S.; Liu, N.; Cui, Y. Manganese Hexacyanomanganate Open Framework as a High-Capacity Positive Electrode Material for Sodium-Ion Batteries. *Nat. Commun.* **2014**, *5*, 5280. [CrossRef]
20. Song, J.; Wang, L.; Lu, Y.; Liu, J.; Guo, B.; Xiao, P.; Lee, J.-J.; Yang, X.-Q.; Henkelman, G.; Goodenough, J.B. Removal of Interstitial H_2O in Hexacyanometallates for a Superior Cathode of a Sodium-Ion Battery. *J. Am. Chem. Soc.* **2015**, *137*, 2658–2664. [CrossRef]
21. Ren, W.; Qin, M.; Zhu, Z.; Yan, M.; Li, Q.; Zhang, L.; Liu, D.; Mai, L. Activation of Sodium Storage Sites in Prussian Blue Analogues via Surface Etching. *Nano Lett.* **2017**, *17*, 4713–4718. [CrossRef] [PubMed]
22. Yue, Y.; Binder, A.J.; Guo, B.; Zhang, Z.; Qiao, Z.-A.; Tian, C.; Dai, S. Mesoporous Prussian Blue Analogues: Template-Free Synthesis and Sodium-Ion Battery Applications. *Angew. Chem. Int. Ed.* **2014**, *53*, 3134–3137. [CrossRef]
23. Wang, L.; Song, J.; Qiao, R.; Wray, L.A.; Hossain, M.A.; Chuang, Y.-D.; Yang, W.; Lu, Y.; Evans, D.; Lee, J.-J.; et al. Rhombohedral Prussian White as Cathode for Rechargeable Sodium-Ion Batteries. *J. Am. Chem. Soc.* **2015**, *137*, 2548–2554. [CrossRef]
24. Pang, W.K.; Kalluri, S.; Peterson, V.K.; Sharma, N.; Kimpton, J.; Johannessen, B.; Liu, H.K.; Dou, S.X.; Guo, Z. Interplay between Electrochemistry and Phase Evolution of the P2-Type $Na_x(Fe_{1/2}Mn_{1/2})O_2$ Cathode for Use in Sodium-Ion Batteries. *Chem. Mater.* **2015**, *27*, 3150–3158. [CrossRef]
25. Zhou, Y.-N.; Wang, P.-F.; Niu, Y.-B.; Li, Q.; Yu, X.; Yin, Y.-X.; Xu, S.; Guo, Y.-G. A P2/P3 Composite Layered Cathode for High-Performance Na-Ion Full Batteries. *Nano Energy* **2019**, *55*, 143–150. [CrossRef]
26. Xie, Y.; Wang, H.; Xu, G.; Wang, J.; Sheng, H.; Chen, Z.; Ren, Y.; Sun, C.-J.; Wen, J.; Wang, J.; et al. In Operando XRD and TXM Study on the Metastable Structure Change of $NaNi_{1/3}Fe_{1/3}Mn_{1/3}O_2$ under Electrochemical Sodium-Ion Intercalation. *Adv. Energy Mater.* **2016**, *6*, 1601306. [CrossRef]
27. Sun, H.-H.; Hwang, J.-Y.; Yoon, C.S.; Heller, A.; Mullins, C.B. Capacity Degradation Mechanism and Cycling Stability Enhancement of AlF_3-Coated Nanorod Gradient $Na[Ni_{0.65}Co_{0.08}Mn_{0.27}]O_2$ Cathode for Sodium-Ion Batteries. *ACS Nano* **2018**, *12*, 12912–12922. [CrossRef]
28. Wei, F.; Zhang, Q.; Zhang, P.; Tian, W.; Dai, K.; Zhang, L.; Mao, J.; Shao, G. Review—Research Progress on Layered Transition Metal Oxide Cathode Materials for Sodium Ion Batteries. *J. Electrochem. Soc.* **2021**, *168*, 050524. [CrossRef]

29. Liu, J.; Zhang, L.-L.; Cao, X.-Z.; Lin, X.; Shen, Y.; Zhang, P.; Wei, C.; Huang, Y.-Y.; Luo, W.; Yang, X.-L. Achieving the Stable Structure and Superior Performance of $Na_3V_2(PO_4)_2O_2F$ Cathodes via Na-Site Regulation. *ACS Appl. Energy Mater.* **2020**, *3*, 7649–7658. [CrossRef]
30. Chao, D.; Lai, C.-H.; Liang, P.; Wei, Q.; Wang, Y.-S.; Zhu, C.; Deng, G.; Doan-Nguyen, V.V.T.; Lin, J.; Mai, L.; et al. Sodium Vanadium Fluorophosphates (NVOPF) Array Cathode Designed for High-Rate Full Sodium Ion Storage Device. *Adv. Energy Mater.* **2018**, *8*, 1800058. [CrossRef]
31. Rajagopalan, R.; Chen, B.; Zhang, Z.; Wu, X.-L.; Du, Y.; Huang, Y.; Li, B.; Zong, Y.; Wang, J.; Nam, G.-H.; et al. Improved Reversibility of Fe^{3+}/Fe^{4+} Redox Couple in Sodium Super Ion Conductor Type $Na_3Fe_2(PO_4)_3$ for Sodium-Ion Batteries. *Adv. Mater.* **2017**, *29*, 1605694. [CrossRef] [PubMed]
32. Wang, X.; Niu, C.; Meng, J.; Hu, P.; Xu, X.; Wei, X.; Zhou, L.; Zhao, K.; Luo, W.; Yan, M.; et al. Novel $K_3V_2(PO_4)_3$/C Bundled Nanowires as Superior Sodium-Ion Battery Electrode with Ultrahigh Cycling Stability. *Adv. Energy Mater.* **2015**, *5*, 1500716. [CrossRef]
33. Park, Y.-U.; Seo, D.-H.; Kwon, H.-S.; Kim, B.; Kim, J.; Kim, H.; Kim, I.; Yoo, H.-I.; Kang, K. A New High-Energy Cathode for a Na-Ion Battery with Ultrahigh Stability. *J. Am. Chem. Soc.* **2013**, *135*, 13870–13878. [CrossRef] [PubMed]
34. Senguttuvan, P.; Rousse, G.; Arroyo y de Dompablo, M.E.; Vezin, H.; Tarascon, J.-M.; Palacín, M.R. Low-Potential Sodium Insertion in a NASICON-Type Structure through the Ti(III)/Ti(II) Redox Couple. *J. Am. Chem. Soc.* **2013**, *135*, 3897–3903. [CrossRef]
35. Chu, Z.; Yue, C. Core–Shell Structured $Na_3V_2(PO_4)_3$/C Nanocrystals Embedded in Multi-Walled Carbon Nanotubes: A High-Performance Cathode for Sodium-Ion Batteries. *Solid State Ion.* **2016**, *287*, 36–41. [CrossRef]
36. Gao, H.; Goodenough, J.B. An Aqueous Symmetric Sodium-Ion Battery with NASICON-Structured $Na_3MnTi(PO_4)_3$. *Angew. Chem. Int. Ed.* **2016**, *55*, 12768–12772. [CrossRef] [PubMed]
37. Gao, H.; Li, Y.; Park, K.; Goodenough, J.B. Sodium Extraction from NASICON-Structured $Na_3MnTi(PO_4)_3$ through Mn(III)/Mn(II) and Mn(IV)/Mn(III) Redox Couples. *Chem. Mater.* **2016**, *28*, 6553–6559. [CrossRef]
38. Zhu, T.; Hu, P.; Wang, X.; Liu, Z.; Luo, W.; Owusu, K.A.; Cao, W.; Shi, C.; Li, J.; Zhou, L.; et al. Realizing Three-Electron Redox Reactions in NASICON-Structured $Na_3MnTi(PO_4)_3$ for Sodium-Ion Batteries. *Adv. Energy Mater.* **2019**, *9*, 1803436. [CrossRef]
39. Liu, Y.; Rong, X.; Bai, R.; Xiao, R.; Xu, C.; Zhang, C.; Xu, J.; Yin, W.; Zhang, Q.; Liang, X.; et al. Identifying the Intrinsic Anti-Site Defect in Manganese-Rich NASICON-Type Cathodes. *Nat. Energy* **2023**, *8*, 1088–1096. [CrossRef]
40. Zhu, T.; Hu, P.; Cai, C.; Liu, Z.; Hu, G.; Kuang, Q.; Mai, L.; Zhou, L. Dual Carbon Decorated $Na_3MnTi(PO_4)_3$: A High-Energy-Density Cathode Material for Sodium-Ion Batteries. *Nano Energy* **2020**, *70*, 104548. [CrossRef]
41. Liu, J.; Huang, Y.; Zhao, Z.; Ren, W.; Li, Z.; Zou, C.; Zhao, L.; Tang, Z.; Li, X.; Wang, M.; et al. Yeast Template-Derived Multielectron Reaction NASICON Structure $Na_3MnTi(PO_4)_3$ for High-Performance Sodium-Ion Batteries. *ACS Appl. Mater. Interfaces* **2021**, *13*, 58585–58595. [CrossRef] [PubMed]
42. Jiang, Y.; Liang, J.; Song, K.; Chen, K.; Li, X.; Zhao, L.; Dai, C.; Zhang, J.; Wang, Y. Benefits of Vanadium Doping in $Na_3MnTi(PO_4)_3$/C as a Potential Candidate for Sodium-Ion Batteries. *Mater. Chem. Phys.* **2022**, *282*, 125938. [CrossRef]
43. Jiang, Y.; Wang, Y.; Li, X.; Zhang, J.; Chen, K.; Liang, J.; Zhao, L.; Dai, C. Chromium Doped NASICON-Structured $Na_3MnTi(PO_4)_3$/C Cathode for High-Performance Sodium-Ion Batteries. *Colloids Surf. A Physicochem. Eng. Asp.* **2022**, *649*, 129340. [CrossRef]
44. Hu, P.; Zhu, T.; Cai, C.; Wang, X.; Zhang, L.; Mai, L.; Zhou, L. A High-Energy NASICON-Type $Na_{3.2}MnTi_{0.8}V_{0.2}(PO_4)_3$ Cathode Material with Reversible 3.2-Electron Redox Reaction for Sodium-Ion Batteries. *Angew. Chem.* **2023**, *135*, e202219304. [CrossRef]
45. Chen, K.; Shi, Q.; Wang, Y.; Li, X.; Jiang, Y.; Xu, H.; Guo, S.; Zhao, L.; Dai, C. Effects of Scandium Doping on the Electrochemical Performance of Cathode Materials $Na_3MnTi(PO_4)_3$ for Sodium-Ion Batteries. *Colloids Surf. A Physicochem. Eng. Asp.* **2023**, *662*, 130996. [CrossRef]
46. Li, H.; Bai, Y.; Wu, F.; Li, Y.; Wu, C. Budding Willow Branches Shaped $Na_3V_2(PO_4)_3$/C Nanofibers Synthesized via an Electrospinning Technique and Used as Cathode Material for Sodium Ion Batteries. *J. Power Sources* **2015**, *273*, 784–792. [CrossRef]
47. Liu, Y.; Wang, F.; Fan, L.-Z. Self-Standing Na-Storage Anode of Fe_2O_3 Nanodots Encapsulated in Porous N-Doped Carbon Nanofibers with Ultra-High Cyclic Stability. *Nano Res.* **2018**, *11*, 4026–4037. [CrossRef]
48. Jin, T.; Liu, Y.; Li, Y.; Cao, K.; Wang, X.; Jiao, L. Electrospun $NaVPO_4F$/C Nanofibers as Self-Standing Cathode Material for Ultralong Cycle Life Na-Ion Batteries. *Adv. Energy Mater.* **2017**, *7*, 1700087. [CrossRef]
49. Conti, D.M.; Fusaro, C.; Bruni, G.; Galinetto, P.; Albini, B.; Milanese, C.; Berbenni, V.; Capsoni, D. ZnS–rGO/CNF Free-Standing Anodes for SIBs: Improved Electrochemical Performance at High C-Rate. *Nanomaterials* **2023**, *13*, 1160. [CrossRef]
50. Yu, S.; Liu, Z.; Tempel, H.; Kungl, H.; Eichel, R.-A. Self-Standing NASICON-Type Electrodes with High Mass Loading for Fast-Cycling All-Phosphate Sodium-Ion Batteries. *J. Mater. Chem. A* **2018**, *6*, 18304–18317. [CrossRef]
51. Meligrana, G.; Ferrari, S.; Lucherini, L.; Celè, J.; Colò, F.; Brugger, J.; Ricciardi, C.; Ruffo, R.; Gerbaldi, C. $Na_3V_2(PO_4)_3$-Supported Electrospun Carbon Nanofiber Nonwoven Fabric as Self-Standing Na-Ion Cell Cathode. *ChemElectroChem* **2020**, *7*, 1652–1659. [CrossRef]
52. Zhang, B.; Yu, Y.; Xu, Z.-L.; Abouali, S.; Akbari, M.; He, Y.-B.; Kang, F.; Kim, J.-K. Correlation Between Atomic Structure and Electrochemical Performance of Anodes Made from Electrospun Carbon Nanofiber Films. *Adv. Energy Mater.* **2014**, *4*, 1301448. [CrossRef]

53. Zhou, Y.; Shao, X.; Lam, K.; Zheng, Y.; Zhao, L.; Wang, K.; Zhao, J.; Chen, F.; Hou, X. Symmetric Sodium-Ion Battery Based on Dual-Electron Reactions of NASICON-Structured $Na_3MnTi(PO_4)_3$ Material. *ACS Appl. Mater. Interfaces* **2020**, *12*, 30328–30335. [CrossRef] [PubMed]
54. *Bruker AXS TOPAS V3.0: General Profile and Structural Analysis Software for Powder Diffraction Data*; User Manual Bruker AXS: Karlsruhe, Germany, 2005.
55. Liu, J.; Huang, Y.; Chen, J.; Fu, L.; Zhao, L.; Zou, C.; Li, X.; Wang, M.; Lin, Y.; Cao, H. Architecture-Improved and Kinetics-Enhanced Nasicon-Type $Na_3MnTi(PO_4)_3$ through Microbe Assisted and Structure Optimized for Boosting Sodium Ion Storage. Available at SSRN. 2022. [CrossRef]
56. Snarskis, G.; Pilipavičius, J.; Gryaznov, D.; Mikoliūnaitė, L.; Vilčiauskas, L. Peculiarities of Phase Formation in Mn-Based Na SuperIonic Conductor (NASICON) Systems: The Case of $Na_{1+2x}Mn_xTi_{2-x}(PO_4)_3$ ($0.0 \leq x \leq 1.5$). *Chem. Mater.* **2021**, *33*, 8394–8403. [CrossRef] [PubMed]
57. Pimenta, M.A.; Dresselhaus, G.; Dresselhaus, M.S.; Cançado, L.G.; Jorio, A.; Saito, R. Studying Disorder in Graphite-Based Systems by Raman Spectroscopy. *Phys. Chem. Chem. Phys.* **2007**, *9*, 1276–1290. [CrossRef] [PubMed]
58. Vu, A.; Qian, Y.; Stein, A. Porous Electrode Materials for Lithium-Ion Batteries—How to Prepare Them and What Makes Them Special. *Adv. Energy Mater.* **2012**, *2*, 1056–1085. [CrossRef]
59. Habedank, J.B.; Günter, F.J.; Billot, N.; Gilles, R.; Neuwirth, T.; Reinhart, G.; Zaeh, M.F. Rapid Electrolyte Wetting of Lithium-Ion Batteries Containing Laser Structured Electrodes: In Situ Visualization by Neutron Radiography. *Int. J. Adv. Manuf. Technol.* **2019**, *102*, 2769–2778. [CrossRef]

nanomaterials

MDPI

Article

Exploring Zinc-Doped Manganese Hexacyanoferrate as Cathode for Aqueous Zinc-Ion Batteries

Julen Beitia [1], Isabel Ahedo [1], Juan Ignacio Paredes [2], Eider Goikolea [1] and Idoia Ruiz de Larramendi [1,*]

[1] Departamento de Química Orgánica e Inorgánica, Universidad del País Vasco (UPV/EHU), Barrio Sarriena s/n, 48940 Leioa, Spain; julen.beitia@ehu.eus (J.B.); isabel.ahedo@ehu.eus (I.A.); eider.goikolea@ehu.eus (E.G.)

[2] Instituto de Ciencia y Tecnología del Carbono, INCAR-CSIC, C/Francisco Pintado Fe 26, 33011 Oviedo, Spain; paredes@incar.csic.es

* Correspondence: idoia.ruizdelarramendi@ehu.eus

Abstract: Aqueous zinc-ion batteries (AZiBs) have emerged as a promising alternative to lithium-ion batteries as energy storage systems from renewable sources. Manganese hexacyanoferrate (MnHCF) is a Prussian Blue analogue that exhibits the ability to insert divalent ions such as Zn^{2+}. However, in an aqueous environment, MnHCF presents weak structural stability and suffers from manganese dissolution. In this work, zinc doping is explored as a strategy to provide the structure with higher stability. Thus, through a simple and easy-to-implement approach, it has been possible to improve the stability and capacity retention of the cathode, although at the expense of reducing the specific capacity of the system. By correctly balancing the amount of zinc introduced into the MnHCF it is possible to reach a compromise in which the loss of capacity is not critical, while better cycling stability is obtained.

Keywords: zinc; Prussian Blue analogue; cathode; aqueous battery

1. Introduction

Nowadays, one of the most critical challenges in our society is to enhance the use of natural and renewable energy sources that are almost inexhaustible and do not generate any environmental degradation. The most important renewable energies are solar, wind, marine, and geothermal. As an advantage, these alternative energy forms allow society to obtain substantial amounts of energy without polluting the environment with greenhouse gas emissions among others. However, these sources depend on geography, climate, and time of day, so it is important to be able to store energy surpluses in an efficient way that allows their use later when the production goes down [1–3]. To overcome the problem related to their intermittent nature, research efforts should also focus on the development of new energy storage systems that help managing the demand/supply dynamics.

Among the various energy storage systems, batteries are particularly important. These systems can store energy in the form of a chemical reaction and later convert it back into electric current. Lithium ion batteries (LIBs) were successfully marketed by SONY in 1991, and ever since then, this battery technology has stood out for its remarkable gravimetric and volumetric energy densities and excellent cycle life compared to other rechargeable batteries. However, this technology has a number of significant drawbacks such as the limited access to lithium reserves, the high cost of the components, and safety concerns related to the flammable nature of organic electrolytes [4]. Moreover, the massive and rapid growth of the electric vehicle market is incentivising the development of alternative technologies to LIBs, not only to meet the needs of the portable energy storage market but also those of stationary applications.

In recent years research has focused on aqueous systems as an alternative to lithium-based systems due to the need for environmentally friendly, safe, and cost-effective devices [5,6]. Therefore, battery systems operating in an aqueous electrolyte are preferable to those working in organic ones. Aqueous electrolytes exhibit interesting advantages such as high ionic conductivity, in addition to being easier to assemble since the need of a controlled atmosphere is eliminated. Among the different aqueous systems, aqueous zinc-ion batteries (AZiBs) stand out for their unique properties linked to the high abundance of metallic zinc, eco-friendliness, intrinsic safety, and cost-effectiveness, making them a promising choice for large-scale renewable energy storage applications [7,8].

AZiBs use metallic zinc in the form of a foil or as powder forming composite electrode, which will oxidize throughout the discharge releasing Zn^{2+} ions. These ions will pass through the near-neutral (or mildly acidic) aqueous electrolyte reaching the cathode where the Zn^{2+} ions will be reversibly inserted into the active material, thanks to the reduction of some electroactive elements that make it up. Throughout the charge, the reverse process occurs, with zinc being electrodeposited on the surface of the anode. During cycling, some issues can occur which are due to inhomogeneous deposition, which can form metallic Zn dendrites and short-circuit the system by the perforation of the separator or the release of hydrogen in the anode (Hydrogen Evolution Reaction, HER) which is formed in the potentials where Zn is electrodeposited. It should be noted that the formation of this gaseous hydrogen is one of the most important issues in the development of AZiBs and needs to be studied in more detail. Recent studies have shown that the control of gas release is possible through control of the pH of the electrolyte or using additives in solution [9,10].

Despite certain issues that still need to be solved, the commercialization of these Zn-ion-based batteries becomes feasible as a suitable, cheap, and environmentally friendly system having high capacity and stability in the long term [11,12]. The selection of a cathode with a high working voltage and a large specific capacity is a critical parameter to increase the gravimetric energy density of AZiBs and to be competitive with the LIBs [13]. In fact, the design and development of cathode materials with high storage capacity, high discharge potential, and a robust crystalline structure with easy insertion and removal pathways has been a great challenge in the development of high-performance AZiBs. Among the most studied materials, oxides such as MnO_2 or V_2O_5 and Prussian Blue analogues (PBAs) are found. Concerning the oxides, MnO_2 presents a good cycle life, but the low ionic and electronic conductivities limit its use, and in the case of vanadium-based oxides, promising capacities and good cycle life are reached but their low operating potential leads to low energy density of the final system [14–16]. The mixed-valence hexacyanoferrate family (Prussian Blue and its analogues) is another group of materials that is of special interest as cathodes in AZiBs. Their open structure, combined with the wide variety of metals that can be part of the structure, allows the electrochemical properties of the material to be tuned [17,18]. These phases have high working potentials, although their specific capacity is limited and they present poor cycling stability [13].

PBAs have attracted attention in the last few years in the field of energy storage due to their easy and inexpensive synthesis procedure by coprecipitation, high specific capacity for the reversible insertion of metallic ions, their high safety and nontoxicity, and the electrochemical properties that can be tuned through the variation of the material composition [19,20]. These materials came from the ancient discovered blue pigment Prussian Blue (PB; $Fe_4[Fe(CN)_6]_3$) which is one of the oldest synthetic coordination polymers. Over the last century, PB has been modified by doping the iron sites with other transition metals such as Mn, Ni, Co, or Zn among others, to obtain analogues (PBAs) and meet different application and research requirements [19,21]. The general chemical formula of PBAs is $A_aTM_A[TM_B(CN)_6]_n \cdot xH_2O$ where A is the alkali metal ion that can be sodium (Na^+) or potassium ions (K^+), while TM_A and TM_B are the transition metal ions which replace iron in PB. In the structure of PBAs (see Figure 1), TM_A is coordinated to the nitrogen atom of the cyanide ion and TM_B is surrounded by the carbon atom of the cyanide ion. This coordination of the cyanide group makes PBAs compounds exhibit a 3D open structure with many interstitial sites where zinc ions can be reversely inserted in the battery cycling process.

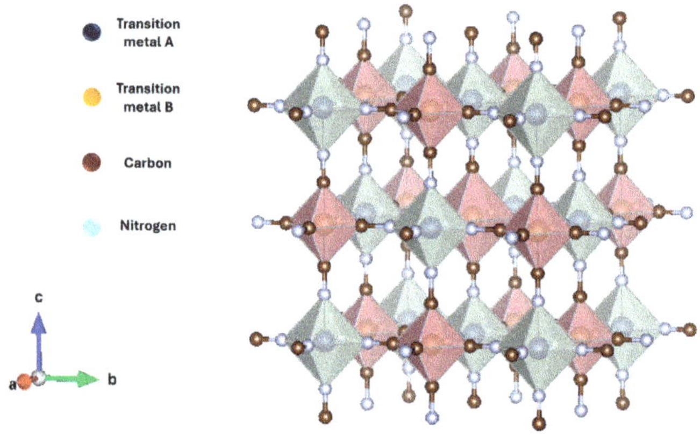

Figure 1. Structure model of a Prussian Blue analogue (PBA).

Among the wide variety of possible PBAs, one of the most studied families is the manganese-based hexacyanoferrates (MnHCF) since it benefits from the existence of two redox couples (Fe^{3+}/Fe^{2+} and Mn^{3+}/Mn^{2+}) delivering large specific capacities [17,18,22]. The charge storage mechanism through Zn^{2+} insertion/extraction in AZiBs using MnHCF as cathodes follows the reaction (1):

$$x\,Zn^{2+} + MnHCF + 2x\,e^{-} \rightleftharpoons Zn_xMnHCF \quad (1)$$

During the discharge, the Zn^{2+} ions produced in the oxidation of the zinc anode are inserted into the MnHCF structure through the reduction of the two active redox couples. During charging, the reverse process occurs, returning the Zn^{2+} ions extracted from the cathode to the anode where they will be reduced to metallic zinc. The amount of zinc that can be reversibly inserted depends on many factors such as water content or particle size, among others. The main problem that MnHCF faces is related to its structural instability in aqueous medium. Although in the first cycles, MnHCF is capable of providing specific capacities as high as 140 mAh g^{-1} (at current densities of 100 mA g^{-1}) [23], the material suffers from structural instability upon cycling, forming new zinc-rich phases and, ultimately, dissolving the cathode material [24]. To avoid these disadvantages, different approaches have been investigated such as the preparation of MnHCFs anchoring MnO_2 [25], the design of hybrid systems such as MnHCF/polymer [23,26], the nanostructuring of the MnHCF through the use of surfactants during the synthesis [27], or adjusting the electrolyte formulation [28,29]. In this work, a different strategy is explored that is based on doping the MnHCF phase with zinc in the manganese position. In this way, the aim is to provide higher structural stability to the material that allows its implementation in AZiBs. This approach allows the preparation of materials through a simple and environmentally friendly synthetic route.

In this work, electrochemical behaviour in Zn chemistry of $K(Mn_{1-x}Zn_x)[Fe(CN)_6]$ (x = 0, 0.25, 0.5, 0.75, and 1) PBAs is studied. This new family of zinc-doped manganese-based PBAs was obtained by coprecipitation through controlling the dropping speed and the reactants' concentration. The structure and morphology of the obtained materials were studied in order to examine the effect of zinc doping. Also, one of the most important goals of this work is to be able to optimize the performance of each of the systems that are used, and testing and finding the most suitable material to achieve greater electrochemical performance. The new family of zinc-doped manganese-based PBAs will be obtained by the coprecipitation method through controlling the dropping speed and the reactants' concentration. The structure and morphology of the obtained materials will be studied in order to examine the effect of zinc doping. Finally, the electrochemical performance of the materials will be examined delving into the role of zinc on the behavior of MnHCF and its structural stability.

2. Materials and Methods

2.1. Synthesis of the PBA Compounds

The synthesis of $K(Mn_{1-x}Zn_x)[Fe(CN)_6]$ (x = 0, 0.25, 0.5, 0.75, and 1), or $Mn_{1-x}Zn_xHCF$, compounds was carried out by coprecipitation method. For the synthesis, stoichiometric quantities of the reagents were mixed as detailed in Table 1. In this regard, $ZnSO_4$ (Sigma-Aldrich, St. Louis, MO, USA) and $Mn(NO_3)_2$ (Sigma-Aldrich, St. Louis, MO, USA) were separately dissolved in distilled water (100 mL) at room temperature. On the other hand, another solution was simultaneously prepared dissolving $K_3Fe(CN)_6$ (0.002 mol, Sigma-Aldrich, St. Louis, MO, USA) in distilled water (100 mL) at room temperature too. Once all the reagents were completely dissolved, PBAs were synthesized by mixing dropwise the solutions (Figure S1), which gave a white-coloured solid in suspension in a solution with a pH of 5–6.

Table 1. Syntheses that were carried out in the work.

Sample	$ZnSO_4$ (mol)	$Mn(NO_3)_2$ (mol)	$K_3Fe(CN)_6$ (mol)	Theoretical Ratio (Zn/Mn)
Zn100	0.002	-	0.002	100:0
Zn75Mn25	0.0015	0.0005	0.002	75:25
Zn50Mn50	0.001	0.001	0.002	50:50
Zn25Mn75	0.0005	0.0015	0.002	25:75
Mn100	-	0.002	0.002	0:100

2.2. Physico-Chemical Characterization of the PBAs

The identification and the structural characterization of the compounds were carried out by X-ray diffraction (XRD, Panalytical X'Pert PRO) between 5 and 70° (2θ) using Cu radiation (Cu-$K_{\alpha a}$ λ = 0.15418 nm). Also, attenuated total reflection infrared spectroscopy (FTIR) was performed using Shimadzu FTIR-8400S (Kyoto, Japan) equipment to analyse directly the pre-cycling materials without any pre-treatment. The morphology of the initial samples was studied by scanning electron microscopy (SEM, Hitachi S-4800, Tokyo, Japan).

2.3. Preparation of Electrodes

The positive electrode composition was 70 wt.% PBA, 20 wt.% conductive Ketjen black carbon, and 10 wt.% polytetrafluoroethylene; PTFE, from a 60 wt.% dispersion in H_2O (Sigma-Aldrich, St. Louis, MO, USA). Negative electrode composition was 99.9 wt.% trace zinc (thickness 0.25 mm, Sigma-Aldrich, St. Louis, MO, USA). This mixture was drenched in ethanol and blended until the plasticity was good enough to obtain a thin black film (150 µm) on carbon paper (H23C6; Quintech Brennstoffzellen Technolgie, Göppingen, Germany). This thin film was left to dry in a vacuum oven at 80 °C for 24 h, and after drying the laminates were cut into 11 mm diameter circular electrodes and finally, weighed and labelled.

2.4. Electrochemical Characterization

The different samples were electrochemically tested using a multichannel potentiostat/galvanostat VMP3 (BioLogic, Seyssinet-Pariset, France) performing cyclic voltammetry measurements at scan rates of 0.1, 0.2, 0.3, 0.5, 0.8, and 1 mV s^{-1}. Galvanostatic charge/discharge measurements using different current densities of 0.1, 0.2, 0.3, 0.5, 1, and 2 A g^{-1} were performed between 0.005–2.0 V vs. Zn^{2+}/Zn.

In the case of the cyclic voltammetry measurements, the assembly of the electrochemical cells was carried out in three-electrode Swagelok® cell systems (Swagelok®, Solon, OH, USA) using stainless steel current collectors, $Mn_{1-x}Zn_xHCF$-based electrodes as working electrodes, metallic zinc disc as the counter, a Ag/AgCl in saturated KCl as the reference electrode, and a porous glass fibre (Whatman GF/A, Maidstone, UK) membrane as the separator with a 3 M zinc trifluoromethanesulfonate ($Zn(OTf)_2$) (Sigma-Aldrich, St. Louis, MO, USA) in distilled water as the electrolyte.

For galvanostatic measurements, $Mn_{1-x}Zn_xHCF$ compounds were tested in two-electrode Swagelok® cell systems (Swagelok®, Solon, OH, USA) using stainless steel current collectors, a metallic zinc disc as both the counter and the reference electrode and a porous glass fibre (Whatman GF/A, Maidstone, UK) membrane as the separator. In all cases, 3 M zinc trifluoromethanesulfonate ($Zn(OTf)_2$) (Sigma-Aldrich, St. Louis, MO, USA) dissolved in distilled water was used as the electrolyte.

3. Results and Discussion

A new family of PBAs with different Zn and Mn contents, $Mn_{1-x}Zn_xHCF$ (x = 0, 0.25, 0.5, 0.75, and 1), has been synthesised and characterised. PBAs can crystallise in different crystal structures depending on factors such as the synthesis process, the type of transition metal, or the degree of zinc insertion. Among the most common structures in the literature cubic, orthorhombic, rhombohedral, and monoclinic are found [30–32]. Figure 2a shows the diffractograms obtained for the synthesised samples together with the theoretical positions of the main diffraction peaks for the cubic (space group: *Fm-3m*) and rhombohedral (space group: *R-3c*) structures. Using the coprecipitation synthesis route, in the case of the manganese sample (Mn100), the diffraction peaks are well-defined and sharp and the diffraction pattern matches with the cubic structure, in good agreement with the structure reported by other authors in the literature [33–37]. The diffraction pattern of the Zn100 sample fits a rhombohedral/hexagonal structure (*R-3c*), as has been described in the bibliography [35,38,39], and no impurities are observable. The samples doped with Zn in the Mn site present a diffraction pattern corresponding to a cubic structure, although as the Zn content increases, the presence of a small peak around 16.2° (2θ), accompanied by other weak signals at higher angles can be detected. This new peak appears at an angle very close to the 100% intensity peak of the rhombohedral phase corresponding to the Zn100 sample. The presence of this peak can be explained based on the presence of a small concentration of a secondary phase with a rhombohedral structure together with a preferential phase with a cubic structure.

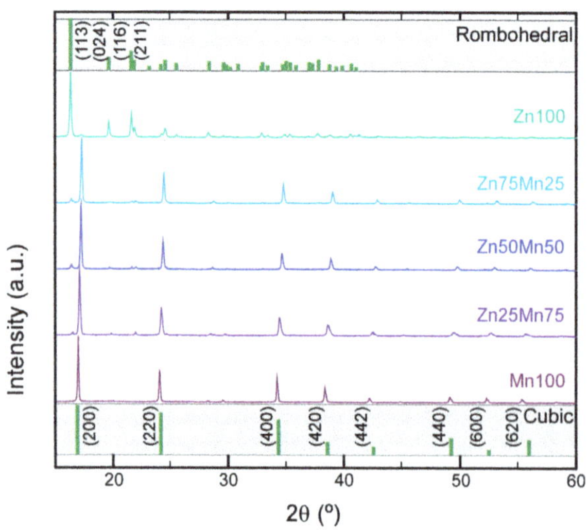

(a)

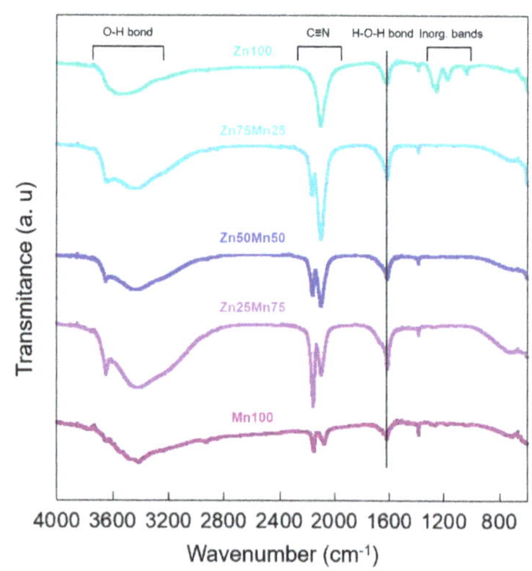

(b)

Figure 2. (**a**) XRD patterns and (**b**) infrared spectra of $Mn_{1-x}Zn_xHCF$ (x = 0, 0.25, 0.5, 0.75, and 1) samples.

In this way, the effect of zinc is evident, distorting the cubic structure that these types of compounds usually exhibit, although zinc-based PBAs with cubic structure have also been reported [40]. The type of structure exhibited by these types of compounds is highly dependent on the synthesis conditions [40,41]. A slight change in the preparation temperature or the pH of the medium can easily cause a change in the structure. In order to avoid this type of effect, all the samples in this work have been synthesised under the same conditions, trying to avoid producing structural or morphological changes beyond those imposed by the change in phase composition.

Going on with the structural study, the analysis and interpretation of the FTIR spectra presented in Figure 2b have been carried out [42–44]. The recorded spectra show one major signal in the 3700–3000 cm^{-1} range related to the H_2O bands (H-O tensile vibration peaks), typical of these compounds [34]. This indicates the presence of water molecules, surely on the surface and even in the open sites of the structure of the compounds. In the central part of the infrared spectrum, between 2300 and 1200 cm^{-1}, four signals can be distinguished. The peaks around 2100 cm^{-1} correspond to the cyanide group (C≡N) and the peak around 1600 cm^{-1} is the H-O-H bond bending vibrations of water molecules [43]. Finally, in the region between 1400 and 1000 cm^{-1}, the bands related to inorganic compounds are observed, and in some cases, in the zone between 1000 and 500 cm^{-1}, the bands corresponding to functional groups bond to the metal can be detected. In this case, two signals that correspond to the bonding of the metal with the cyanide group are observed. From the analysis of the spectra, two important features are highlighted. On the one hand, the existence of signals clearly assignable to water molecules makes clear the presence of water in the compounds. On the other hand, the change in the intensity ratio of the bands attributed to the stretching vibration peak of the bridging cyano group (–CN), suggests a change in the coordination environment of this group depending on the Zn/Mn ratio. In the Zn100 sample, a single peak can be seen that, as the amount of Mn increases, loses intensity, and a new peak appears at slightly higher wavenumber values (blue shift). The positions of the bands corresponding to the cyano group assigned to the Fe^{III}—CN—M^{II} and Fe^{II}—CN—M^{II} chains are around 2160 and 2080 cm^{-1}, respectively, for the samples that have manganese in their composition. In the case of the Zn100 phase, a single band relative to the Fe^{II}-C≡N-M^{III} bonds is observed, located at 2100 cm^{-1} [45,46]. The shift appreciated in the bands is due to the change in the electronic state of the cations linked to the N, which undergo a transition from high-spin (MII) to low-spin (MIII) [47].

The effect of Zn doping at the Mn site on the morphology of the materials was analyzed by SEM and the recorded images are shown in Figure 3. The Mn100 sample exhibits a cubic shape with particles of different sizes, between 50 nm and 3 μm, as has been widely reported in the literature [26,33,48,49]. The solubility constant of the $K_2Mn[Fe(CN)_6]$ phase (K_{sp} = $10^{-12.1}$) is very low [50], which causes this compound to precipitate very easily and quickly in solution, giving rise to particles without a defined shape. When the precipitation process is controlled by adding, for example, chelating agents or slowing down the formation of the product, it is possible to obtain cubic-shaped particles with few defects [51]. In the synthesis process of the PBAs presented in this work, special care has been taken to control the reaction time, adding the reagents simultaneously and slowly, drop by drop, in order to control the nucleation and growth of the particles in an efficient and simple way. In this case, it is critical to control the synthesis conditions since they have a great impact both on the structure of the compound and on the size and morphology of the particles. Lee and Huh already demonstrated that by increasing the concentration of HNO_3 in the synthesis of $KFe^{III}[Fe^{II}(CN)_6$, the particle morphology evolved from cubes to star-like hexapods [41]. This morphological change allowed them to determine that the oxidation processes began in the corners of the cubes, which act as active sites. Zhang et al. carried out a similar study on zinc hexacyanoferrate, and by adjusting the concentration of reagents and controlling the dropping speed, particles with different polyhedral shapes were prepared: cuboctahedrons, truncated octahedron, and octahedron zinc hexacyanoferrates [40]. In fact, it has been reported that in the case of certain zinc-based PBAs, the cubic structure is not stable, since water molecules can leave the structure relatively easily, giving rise to a transformation towards a rhombohedral phase [52,53]. In our case, Zn100 particles show a cuboctahedral shape, very different from that of Mn100 particles. Furthermore, the particles appear to be covered with small fragments, presenting rough surfaces, unlike the Mn100 sample in which the surfaces of the sides of the cubes are quite smooth. As the zinc ratio of the samples increases, the shape of the particles evolves from cubes to cuboctahedrons. As determined in the structural study, the samples tend to maintain the structure and shape imposed by manganese in the PBAs, but at high Zn contents, the compound is forced to readapt, taking on more importance the characteristics of the Zn100 compound. Thus, while the Zn25Mn75 sample is made up of cubic particles with an edge of around 100 nm, the particles of the Zn75Mn25 sample show a clearly cuboctahedral shape, with smooth and clean surfaces.

Figure 3. SEM images of $Mn_{1-x}Zn_xHCF$ (x = 0, 0.25, 0.5, 0.75 and 1) samples.

Indeed, the composition of the phases is vital to determine the electrochemical properties of the material, but it has been shown that the morphology of the particles also has an important impact on the performance of the cathode. Specifically, the cuboctahedral shape of the Zn-based PBAs provides larger discharge capacities at high rates, in addition to exhibiting greater stability in cycling [40]. This fact has been related to the high surface area that this shape of the particles presents. In the case of the compounds in this study, a change in the morphology of the particles is observed, although there does not seem to be an effect on the particle size with the zinc content.

Before integrating the prepared materials into AZiBs, the reaction kinetics of each compound were studied by means of cyclic voltammetry (CV) measurements at different scan rates from 0.1 to 1.0 mV s^{-1}. It should be noted that in the case of Zn100 sample, the CV obtained does not show any peak (Figure S2) and therefore its electrochemical activity is practically zero, in good agreement with results reported by other authors [54,55]. Also, as can be seen in Figure 4, in the case samples with high amounts of Mn (Mn100 and Zn25Mn75), CV curves show two major peaks, which maintain the shape during the experiment at different rates. The peaks that appear at 0.95/0.98 V vs. Ag/AgCl during oxidation and at ca. 0.90 V during reduction correspond to the Mn^{3+}/Mn^{2+} redox pair and, as expected, as the scan rate increases the intensity of the peak increases, perceiving a slight shift towards higher potentials in the oxidation branch and towards lower values in the reduction branch. By increasing the zinc content (and therefore decreasing the amount of Mn in the sample), the appearance of another signal at 0.9/0.87 V vs. Ag/AgCl (oxidation) and ca. 0.75 V (reduction) is distinguished for the Zn50Mn50 phase that is assigned to the Fe^{3+}/Fe^{2+} redox pair [39,56]. The intensity ratio of the redox pairs corresponding to Mn and Fe is further balanced when examining the sample Zn75Mn25. Taking into account the maximum intensities reached by the different samples, the superiority in terms of electrochemical activity of the Mn100 sample is clear.

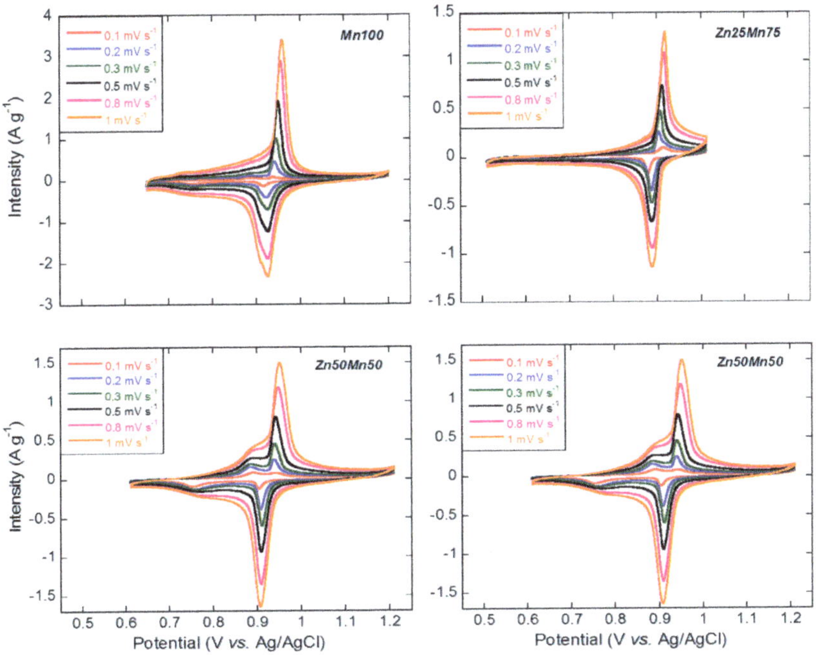

Figure 4. CV curves of the $Mn_{1-x}Zn_xHCF$ (x = 0, 0.25, 0.5 and 0.75) samples at different scan rates: 0.1, 0.2, 0.3, 0.5, 0.8 and 1 mV s^{-1}.

In Figure S3 the relationship between the peak current (I_p; mA g^{-1}) and the square root of the used scan rates ($\sqrt{V}$; $\sqrt{mVs^{-1}}$) is represented [57]. This relationship is almost linear in all the samples, which indicates that all systems are compatible with and ideal diffusion-controlled faradaic process. Furthermore, the anodic and cathodic lines present great symmetry, which is indicative of the high reversibility of the redox process. Moreover, the peak separation at 0.1 mV s^{-1} is about 20 mV and increases just up to 35 mV at 1 mV s^{-1}, which is also indicative of the reversibility of the reaction.

Galvanostatic charge/discharge measurements were performed between 0.005–2.0 V vs. Zn^{2+}/Zn using a current density of 0.1 A g^{-1}. Figure S4 shows the evolution of the charge/discharge curves over 10 cycles for each sample. Again, it is possible to distinguish two different behaviours depending on the Zn content in the sample. On the one hand, the samples rich in manganese, Mn100 and Zn25Mn75, show 2–3 plateaus, being more defined in the case of the Mn100 sample. These small plateaus have been also reported in previous works using PBAs, specifically in works in which manganese has been used in this type of structure [58–60]. In the first discharge curves of the two samples with the highest Mn content, a small plateau is distinguished around 1.8 V vs. Zn^{2+}/Zn that progressively disappears, being imperceptible from the seventh cycle onwards. Likewise, two other plateaus are distinguished between 0.8 and 1.5 V, the one with the highest potential being assigned to the $Mn^{3+/2+}$ redox couple, while the one at lower potentials corresponds to the $Fe^{3+/2+}$ redox couple, as was established in the analysis of CVs. The contribution to the capacity of both processes decreases upon cycling, with the plateau due to $Fe^{3+/2+}$ disappearing in the last cycles. This change in the profiles, which ultimately become sloppy, may be related to structural changes that are usually related to the dissolution of Mn, leading to the formation of low crystalline MnO_2 [61], and changes in the coordination environment of the transition metals in the structure [60]. When a higher amount of Zn is introduced into the structure (Zn50Mn50 and Zn75Mn25 samples), it is barely possible to distinguish the presence of well-defined plateaus; instead, a progressive drop is perceived. This type of profile differs from those reported by other authors analyzing zinc hexacyanoferrates in which only iron acts as an electrochemically active transition metal, distinguishing a plateau corresponding to the transference of one electron in sodium-based batteries [35]. Here, the presence of iron accompanied by manganese, which also presents electrochemical activity, produces a clear change in the profile. In any case, similar discharge profiles have been reported for zinc-doped hexacyanoferrates [26]. The profiles in-the-form slope in AZiBs have been justified based on the (in)stability of the PBAs, especially in the first cycles in which a structural readjustment occurs due to the deintercalation of the K^+ cations [56]. Kim et al. analyzed the effect of the introduction of Zn in cobalt hexacyanoferrates, observing a similar transition from a cubic structure of the Co phase to a rhombohedral one as the Zn content increased [62]. The coexistence of both phases with different structures in the samples doped with Zn (mixture of cubic and rhombohedral phases observed in the study using XRD), together with the decrease in the content of electrochemically active metals causes a decrease in capacity, leading to a loss of activity observed in the case of the Zn100 phase. The presence of the rhombohedral phase in the samples with Zn facilitates higher stability in these compounds since the structural changes due to the (de)intercalation of Zn^{2+} ions in the cyclability are less severe [62]. This way, the stability of the material is higher, allowing for superior capacity retention.

In terms of electrochemical properties, the higher the Mn content is, the higher the capacity values are, as shown in Figure 5. In contrast, at high Mn contents, capacity values decrease significantly by the tenth cycle. For instance, for Mn100 sample capacity retention is 39% (130 mA h g^{-1} vs. 51 mA h g^{-1}) while for Zn25Mn75 sample this retention is 53% (98 mA h g^{-1} vs. 52 mA h g^{-1}). For Zn-enriched samples, instead, the capacity retention is much higher: 61% for Zn50Mn50 (64 mA h g^{-1} vs. 39 mA h g^{-1}) and 94.5% for Zn75Mn25 (73 mA h g^{-1} vs. 69 mA h g^{-1}). At low current intensities, the active material has enough time to fully charge and discharge, i.e., Zn^{2+} ions can intercalate and deintercalate into the PBA structure. As the current increases, diffusion processes and even charge transfer processes are hindered and the reaction yield is much lower. On the other hand, the decrease in capacity with increasing zinc content is related to the lack of electroactivity of Zn. The higher the concentration of Zn in the PBA structure, the lower the amount of $Mn^{3+/2+}$ redox couple, which has a direct impact on the capacity of the compound.

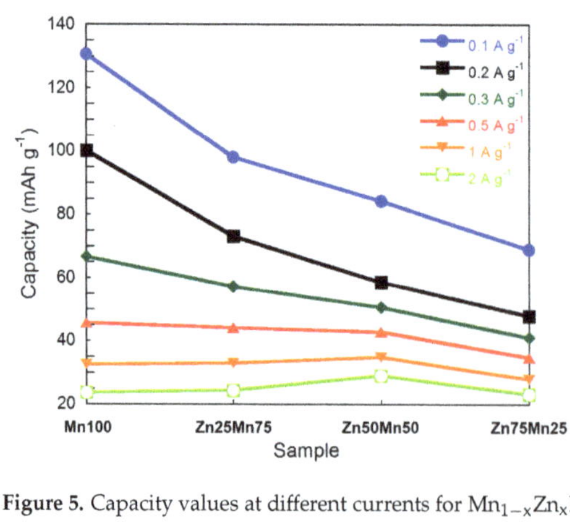

Figure 5. Capacity values at different currents for $Mn_{1-x}Zn_xHCF$ (x = 0, 0.25, 0.5, and 0.75) samples.

Galvanostatic discharge and charge profiles at different current densities, i.e., 0.1, 0.2, 0.3, 0.5, 1.0, and 2.0 A g^{-1} are shown in Figure S5. As has been pointed out, at high current densities the capacity of the system decays, due to the limitations imposed in diffusion and charge transfer processes [63]. These limitations also cause the polarization to grow. This increase in polarization is more evident in samples with higher Mn content, which can be attributed to a higher difficulty of the cubic structure to facilitate the diffusion of Zn^{2+} ions. The comparison of the rate performance for the different materials analyzed in this work is shown in Figure 6.

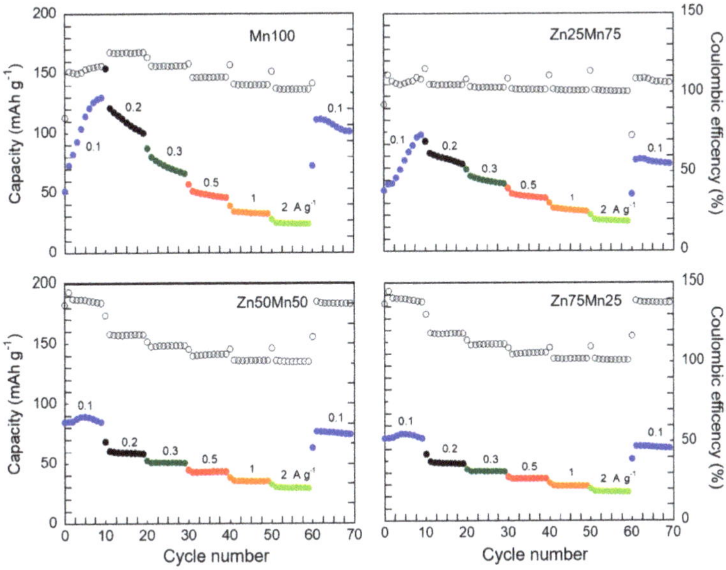

Figure 6. Rate performance and Coulombic efficiencies (open dots) of $Mn_{1-x}Zn_xHCF$ (x = 0, 0.25, 0.5 and 0.75) samples.

In terms of capacity, the values delivered by the Mn100 sample are higher than in the case of the Zn-doped phases, again, which is related to the higher amount of electrochemically active metals in Mn-rich phases. An increase in capacity is observed in the first cycles at 0.1 A g^{-1}, being more critical for samples with higher Mn content. This effect in PBAs applied to AZiBs has already been reported in the literature [27,56] and is related to the structural readjustment of this type of compounds due to the need to deintercalate K^+ cations. This process occurs in the first cycles, facilitating higher intercalation of Zn^{2+} ions as K^+ leaves the structure. This structural conditioning process requires a series of cycles to occur, reaching maximum a capacity around the tenth cycle. Instead, in the case of Zn-enriched PBAs (Zn50Mn50 and Zn75Mn25 samples), the variation in the capacity values in the first cycles is less notable, due to the stability conferred on the structure by the presence of high amounts of zinc. The mechanism associated with this electrode activation process involves the release of K^+ ions from the initial PBA that will be replaced by Zn^{2+} ions throughout the cycles. As has been demonstrated by theoretical calculations [64], in the initial phase due to de presence of K^+, the MnN_6 octahedra presents a moderate distortion of the structure due to the Jahn-Teller effect of trivalent manganese. The introduction of Zn^{2+} ions into the structure during the first cycles induces a much more intense Jahn–Teller effect that causes uncontrolled phase transitions in the material. This effect is more obvious the greater the amount of manganese present in the compound, so as the position of manganese is doped with zinc, the consequences of the structural distortion induced by the Jahn-Teller effect become less intense (since a smaller amount of Mn(III) exists in the compound) until it practically disappears for the Zn75Mn25 phase. Regarding the impact of the current intensity, the loss of capacity experienced by the samples when increasing the current is smaller when increasing the zinc content. In the case of Mn100 and Zn25Mn75, in the first cycles at 0.1 A g^{-1}, the capacity increases considerably, reaching a maximum, that is comparable or even higher to other PBA families [23,25,64,65].

In order to corroborate the effect of zinc doping of the MnHCF phase on structural stability, cycling tests for 200 cycles have been carried out applying a current intensity of 50 mA g^{-1} (Figure 7).

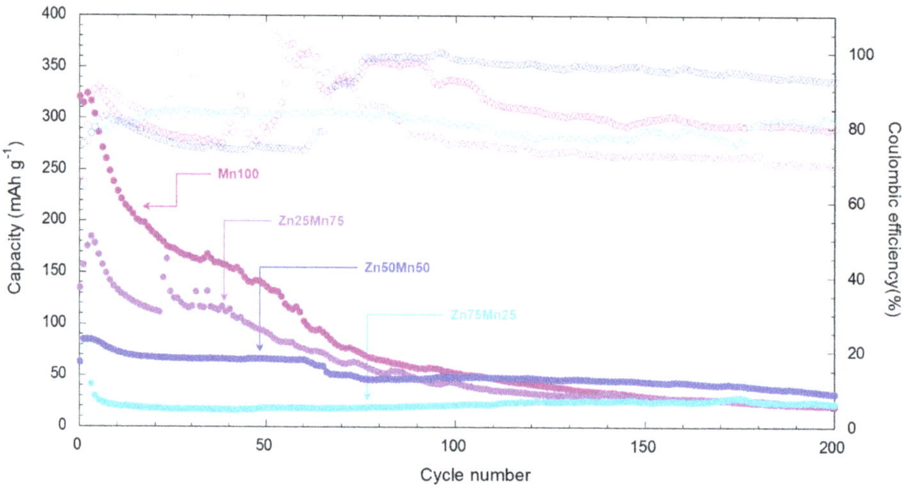

Figure 7. Long-term cycling of $Mn_{1-x}Zn_xHCF$ (x = 0, 0.25, 0.5 and 0.75) samples at 50 mA g^{-1}. The open dots refer to the Coulombic efficiency.

The low initial efficiencies are related to the irreversible insertion of Zn^{2+} ions, as has been reported by other authors [64]. The Mn100 sample presents the highest capacity but by cycle 20 it has already been reduced practically by half (184.7 mAh g^{-1}) and continues to decay steadily until the end of the experiment (21 mAh g^{-1} in cycle 200, which represents only 6.5% of the initial capacity of 323 mAh g^{-1}). As zinc replaces manganese in the MnHCF phase, a lower initial capacity is obtained related to a lower concentration of electrochemically active metals (Mn and Fe), as has been mentioned previously and described by other authors in similar studies [64,66]. In the case of zinc-rich phases, a lower effect of the electrode activation process and greater stability in terms of capacity can be seen. For the Zn50Mn50 sample, the capacity stabilizes at 70 mAh g^{-1} (82.6% retention of initial capacity of 84.7 mAh g^{-1}) starting from cycle 10. This capacity value is maintained until cycles 60–65, where an abrupt drop occurs, up to 48 mAh g^{-1} which represents 56.7% of the initial capacity. The Zn75Mn25 sample suffers a drastic drop in capacity in the first 5 cycles, stabilizing around 23 mAh g^{-1} (55.3% initial capacity retention of 41.6 mAh g^{-1}) throughout the entire cycling test. The manganese-rich samples, Mn100 and Zn25Mn75, exhibit a severe loss of capacity in the first 60–70 cycles, reaching capacities of 21 and 23 mAh g^{-1} in cycle 200, respectively (6.5% and 13.14% initial capacity retention). This significant loss of capacity is related to the dissolution of manganese in the aqueous medium of the electrolyte. Although the MnHCF phase presents great stability in an organic medium, it tends to undergo dissolution reactions of manganese ions in the electrolyte [24]. As a consequence of these processes, δ-MnO_2 is formed, which is capable of reacting in the presence of protons, reducing to MnOOH, which in turn can be reduced in the aqueous medium to produce Mn^{2+} ions [13,67], through the following reactions:

$$MnO_2 + H^+ + e^- \rightarrow MnOOH \quad (2)$$

$$MnOOH + 3\,H^+ + e^- \rightarrow Mn^{2+} + 2\,H_2O \quad (3)$$

The loss of manganese ions in the PBA produces the formation of a zinc-rich phase that has a structure similar to that of the $K_2Zn_3[Fe(CN)_6]_2$ phase. In order to delve deeper into the stability of the compounds, a study using X-ray diffraction and scanning electron microscopy has been carried out on the electrodes once cycled. The SEM images and the diffractograms recorded after the long-term cycling experiment are shown in Figures S6 and S7. SEM analysis has been carried out to identify changes in morphology in the active material. In the images in Figure S6 it can be seen that the morphology of the materials is similar for all the electrodes, showing the presence of fibres that are remains of the glass fibre separator. Regarding the structural study, the manganese-rich samples (Mn100 and Zn25Mn75) undergo a clear structural change after cycling (Figure S7). The most characteristic peak of the cubic phase disappears and a peak at 24.33° (2θ) emerges with maximum intensity. The new diffraction profile fits with manganese oxohydroxide (MnOOH, JCPDS#41-1379) along with traces of the cubic phase (see Figure S8), which corroborates the dissolution process of manganese throughout the cycling. In this way, the degradation of performance is associated with the loss of active material. This dissolution reaction is less evident as zinc is introduced in the manganese position of the PBA, where the peaks corresponding to the initial compound are more visible. The post-mortem analysis corroborates that the introduction of zinc in the manganese sites of the MnHCF provides greater stability to the structure, preventing the insertion and extraction of Zn^{2+} ions in the battery cycle from inducing a phase transformation in the active material.

The analysis of the electrochemical response of the different materials presented in this work shows that zinc doping of manganese hexacyanoferrate provides greater stability to the material, causing, in turn, a loss in specific capacity. These results indicate that these compounds allow the reversible insertion/extraction of Zn^{2+} ions, without observing degradation upon cycling. Zinc doping is an easy strategy to implement to produce more stable cathode materials, although it is necessary to balance the amount of zinc to avoid a drastic loss of AZiB capacity.

4. Conclusions

In this study, the synthesis, physico-chemical characterisation and electrochemical study of the $Mn_{1-x}Zn_xHCF$ (x = 0, 0.25, 0.5, 0.75 and 1) Prussian Blue analogues as cathodes for rechargeable aqueous Zn-ion batteries has been carried out. Substituting manganese with zinc induces a change in the structure, and a secondary rhombohedral phase appears accompanying the predominant cubic structure. This distortion in the phases has also been corroborated by infrared spectroscopy. The introduction of zinc into the MnHCF structure produces a change in morphology from a cubic to a cuboctahedral shape, which could translate into a larger specific area. This structural and morphological change has a direct impact on the electrochemical response of the compounds. Pristine MnHCF compound provides a high capacity of 130 mAh g^{-1} at 0.1 A g^{-1}, but upon cycling this capacity is quickly lost due to the inherent weak structural stability of the compound. The instability of the Mn-rich phases has been corroborated through post-cycling studies, which demonstrate that manganese is susceptible to dissolving throughout the charge/discharge cycles, giving rise to the formation of the MnOOH compound. The impact of zinc substitution in manganese-based PBAs causes a decrease in the electrochemical activity, which is mainly due to the Mn^{3+}/Mn^{2+} redox couple. Despite decreasing the specific capacity of the system, increasing the amount of zinc improves capacity retention and Coulombic efficiency. This way, it is proven that through a simple approach, by zinc doping, it is possible to obtain higher reversibility and stable performance for the MnHCF phase.

Supplementary Materials: The following supporting information can be downloaded at: https://www.mdpi.com/article/10.3390/nano14131092/s1, Figure S1: Schematic representation of the experimental setup for the synthesis of materials by the co-precipitation method; Figure S2: CV curves of the Zn100 sample at different scan rates: 1, 2, 5 and 10 mV s^{-1}; Figure S3: Randles-Sevcik Plot for the

$Mn_{1-x}Zn_xHCF$ (x = 0, 0.25, 0.5, 0.75) samples; Figure S4: Discharge/charge profiles of the $Mn_{1-x}Zn_xHCF$ (x = 0, 0.25, 0.5, 0.75) materials at a current density of 0.1 A g^{-1} in the 0.005–2.0 V vs. Zn^{2+}/Zn range; Figure S5: Galvanostatic charge–discharge voltage profiles at different applied currents of the $Mn_{1-x}Zn_xHCF$ (x = 0, 0.25, 0.5, 0.75) materials; Figure S6. SEM images of the $Mn_{1-x}Zn_xHCF$ (x = 0, 0.25, 0.5, 0.75) electrodes after cyclability tests: (a,b) Zn75Mn25, (c,d) Zn50Mn50, (e,f) Zn25Mn75, and (g,h) Mn100; Figure S7. XRD patterns of the $Mn_{1-x}Zn_xHCF$ (x = 0, 0.25, 0.5, 0.75) samples recorded after the synthesis (raw material), of the electrode before the electrochemical study (pristine) and of the post-mortem electrode after the cyclability study (post-cycling); Figure S8. Le Bail profile fitting of the powder X-ray diffraction pattern corresponding to the Zn25Mn75 electrode after cycling test.

Author Contributions: Conceptualization, J.I.P. and I.R.d.L.; methodology, E.G., J.I.P. and I.R.d.L.; validation, E.G., J.I.P. and I.R.d.L.; formal analysis, J.B. and E.G.; investigation, J.B. and I.A.; resources, J.B. and I.A.; data curation, J.B., E.G. and I.R.d.L.; writing—original draft preparation, J.B. and I.R.d.L.; writing—review and editing, E.G., J.I.P. and I.R.d.L.; supervision, E.G., J.I.P. and I.R.d.L.; project administration, E.G., J.I.P. and I.R.d.L.; funding acquisition, E.G., J.I.P. and I.R.d.L. All authors have read and agreed to the published version of the manuscript.

Funding: This research was funded by Gobierno Vasco/Eusko Jaurlaritza, project IT1546-22 and by MICIU/AEI/10.13039/501100011033 and by the "European Union NextGenerationEU/PRTR, projects TED2021-131517B-C21 and TED2021-131517B-C22, as well as by MICINN/AEI/10.13039/501-100011033 and the European Regional Development Fund (ERDF, A way of making Europe) through project PID2021-125246OB-I00.

Data Availability Statement: The data presented in the study are available from the corresponding author.

Acknowledgments: The authors thank for technical and human support provided by SGIker (UPV/EHU/ ERDF, EU).

Conflicts of Interest: The authors declare no conflicts of interest.

References

1. Sayed, E.T.; Olabi, A.G.; Alami, A.H.; Radwan, A.; Mdallal, A.; Rezk, A.; Abdelkareem, M.A. Renewable Energy and Energy Storage Systems. *Energies* **2023**, *16*, 1415. [CrossRef]
2. Babuder, M.; Omahen, G.; Bregar, Z. New Age with Renewable Energy Sources. *Elektrotechnik Informationstechnik* **2013**, *130*, 241–246. [CrossRef]
3. Tokar, D.; Foris, D.; Tokar, A.; Foris, T. Sustainable Development of Disadvantaged Regions by Renewable Energy Sources Integration. In Proceedings of the 2019 International Conference ESRD, Jelgava, Letonia, 8 May 2019; pp. 237–244.
4. Goikolea, E.; Palomares, V.; Wang, S.; de Larramendi, I.R.; Guo, X.; Wang, G.; Rojo, T. Na-Ion Batteries—Approaching Old and New Challenges. *Adv. Energy Mater.* **2020**, *10*, 1–21. [CrossRef]
5. Ahn, H.; Kim, D.; Lee, M.; Nam, K.W. Challenges and Possibilities for Aqueous Battery Systems. *Commun. Mater.* **2023**, *4*, 37. [CrossRef]
6. Yi, J.; Xia, Y. Advanced Aqueous Batteries: Status and Challenges. *MRS Energy Sustain.* **2022**, *9*, 106–128. [CrossRef]
7. Zhu, K.; Wu, T.; Sun, S.; Wen, Y.; Huang, K. Electrode Materials for Practical Rechargeable Aqueous Zn-Ion Batteries: Challenges and Opportunities. *ChemElectroChem* **2020**, *7*, 2714–2734. [CrossRef]
8. Hong, X.; Deng, C.; He, J.; Liang, B.; Wang, G.; Tu, Z. Existing Electrochemical Activation Mechanisms and Related Cathode Materials for Aqueous Zn Ion Batteries. *Energy Convers. Manag.* **2024**, *299*, 117906. [CrossRef]
9. Bani Hashemi, A.; Kasiri, G.; La Mantia, F. The Effect of Polyethyleneimine as an Electrolyte Additive on Zinc Electrodeposition Mechanism in Aqueous Zinc-Ion Batteries. *Electrochim. Acta* **2017**, *258*, 703–708. [CrossRef]
10. Higashi, S.; Lee, S.W.; Lee, J.S.; Takechi, K.; Cui, Y. Avoiding Short Circuits from Zinc Metal Dendrites in Anode by Backside-Plating Configuration. *Nat. Commun.* **2016**, *7*, 11801. [CrossRef]
11. Zampardi, G.; La Mantia, F. Open Challenges and Good Experimental Practices in the Research Field of Aqueous Zn-Ion Batteries. *Nat. Commun.* **2022**, *13*, 687. [CrossRef]
12. Li, X.; Wang, L.; Fu, Y.; Dang, H.; Wang, D.; Ran, F. Optimization Strategies toward Advanced Aqueous Zinc-Ion Batteries: From Facing Key Issues to Viable Solutions. *Nano Energy* **2023**, *116*, 108858. [CrossRef]
13. Jia, X.; Liu, C.; Neale, Z.G.; Yang, J.; Cao, G. Active Materials for Aqueous Zinc Ion Batteries: Synthesis, Crystal Structure, Morphology, and Electrochemistry. *Chem. Rev.* **2020**, *120*, 7795–7866. [CrossRef]
14. Wu, Y.; Fee, J.; Tobin, Z.; Shirazi-Amin, A.; Kerns, P.; Dissanayake, S.; Mirich, A.; Suib, S.L. Amorphous Manganese Oxides: An Approach for Reversible Aqueous Zinc-Ion Batteries. *ACS Appl. Energy Mater.* **2020**, *3*, 1627–1633. [CrossRef]
15. Dai, Q.; Li, L.; Hoang, T.K.A.; Tu, T.; Hu, B.; Jia, Y.; Zhang, M.; Song, L.; Trudeau, M.L. The Secondary Aqueous Zinc-Manganese Battery. *J. Energy Storage* **2022**, *55*, 105397. [CrossRef]

16. Senguttuvan, P.; Han, S.D.; Kim, S.; Lipson, A.L.; Tepavcevic, S.; Fister, T.T.; Bloom, I.D.; Burrell, A.K.; Johnson, C.S. A High Power Rechargeable Nonaqueous Multivalent Zn/V_2O_5 Battery. *Adv. Energy Mater.* **2016**, *6*, 1600826. [CrossRef]
17. Zhou, A.; Cheng, W.; Wang, W.; Zhao, Q.; Xie, J.; Zhang, W.; Gao, H.; Xue, L.; Li, J. Hexacyanoferrate-Type Prussian Blue Analogs: Principles and Advances Toward High-Performance Sodium and Potassium Ion Batteries. *Adv. Energy Mater.* **2021**, *11*, 1–35. [CrossRef]
18. Paolella, A.; Faure, C.; Timoshevskii, V.; Marras, S.; Bertoni, G.; Guerfi, A.; Vijh, A.; Armand, M.; Zaghib, K. A Review on Hexacyanoferrate-Based Materials for Energy Storage and Smart Windows: Challenges and Perspectives. *J. Mater. Chem. A Mater.* **2017**, *5*, 18919–18932. [CrossRef]
19. Li, W.J.; Han, C.; Cheng, G.; Chou, S.L.; Liu, H.K.; Dou, S.X. Chemical Properties, Structural Properties, and Energy Storage Applications of Prussian Blue Analogues. *Small* **2019**, *15*, 1900470. [CrossRef]
20. Zeng, Y.; Lu, X.F.; Zhang, S.L.; Luan, D.; Li, S.; Lou, X.W. Construction of Co–Mn Prussian Blue Analog Hollow Spheres for Efficient Aqueous Zn-Ion Batteries. *Angew. Chem. Int. Ed.* **2021**, *60*, 22189–22194. [CrossRef]
21. Xu, C.; Yang, Z.; Zhang, X.; Xia, M.; Yan, H.; Li, J.; Yu, H.; Zhang, L.; Shu, J. Correction to: Prussian Blue Analogues in Aqueous Batteries and Desalination Batteries. *Nanomicro Lett.* **2021**, *13*, 187. [CrossRef]
22. Wang, D.; Lv, H.; Hussain, T.; Yang, Q.; Liang, G.; Zhao, Y.; Ma, L.; Li, Q.; Li, H.; Dong, B.; et al. A Manganese Hexacyanoferrate Framework with Enlarged Ion Tunnels and Two-Species Redox Reaction for Aqueous Al-Ion Batteries. *Nano Energy* **2021**, *84*, 105945. [CrossRef]
23. Cao, T.; Zhang, F.; Chen, M.; Shao, T.; Li, Z.; Xu, Q.; Cheng, D.; Liu, H.; Xia, Y. Cubic Manganese Potassium Hexacyanoferrate Regulated by Controlling of the Water and Defects as a High-Capacity and Stable Cathode Material for Rechargeable Aqueous Zinc-Ion Batteries. *ACS Appl. Mater. Interfaces* **2021**, *13*, 26924–26935. [CrossRef] [PubMed]
24. Li, M.; Maisuradze, M.; Sciacca, R.; Hasa, I.; Giorgetti, M. A Structural Perspective on Prussian Blue Analogues for Aqueous Zinc-Ion Batteries. *Batter. Supercaps* **2023**, *6*, e202300340. [CrossRef]
25. Chen, J.; Liao, L.; Sun, B.; Song, X.; Wang, M.; Guo, B.; Ma, Z.; Yu, B.; Li, X. Manganese Hexacyanoferrate Anchoring MnO_2 with Enhanced Stability for Aqueous Zinc-Ion Batteries. *J. Alloys Compd.* **2022**, *903*, 163833. [CrossRef]
26. Dai, P.; Ma, W.; Zhou, Y.; Tang, Y.; Cao, X.; Wu, P.; Zhou, H. Superfast Phase Transformation Driven by Dual Chemical Equilibrium Enabling Enhanced Electrochemical Energy Storage. *Adv. Funct. Mater.* **2024**, *2308357*, 1–8. [CrossRef]
27. Liu, B.; Zhao, R.; Zhang, Q.; Ali, U.; Li, Y.; Hao, Y.; Jia, H.; Li, Y.; Zeng, G.; Sun, M.; et al. Ultrafine Manganese Hexacyanoferrate with Low Defects Regulated by Potassium Polyacrylate for High-Performance Aqueous Zn-Ion Batteries. *J. Energy Storage* **2023**, *72*, 108535. [CrossRef]
28. Wang, Y.; Li, Q.; Hong, H.; Yang, S.; Zhang, R.; Wang, X.; Jin, X.; Xiong, B.; Bai, S.; Zhi, C. Lean-Water Hydrogel Electrolyte for Zinc Ion Batteries. *Nat. Commun.* **2023**, *14*, 3890. [CrossRef] [PubMed]
29. Luo, L.; Liu, Y.; Shen, Z.; Wen, Z.; Chen, S.; Hong, G. High-Voltage and Stable Manganese Hexacyanoferrate/Zinc Batteries Using Gel Electrolytes. *ACS Appl. Mater. Interfaces* **2023**, *15*, 29032–29041. [CrossRef] [PubMed]
30. Peng, F.; Yu, L.; Gao, P.; Liao, X.Z.; Wen, J.; He, Y.S.; Tan, G.; Ren, Y.; Ma, Z.F. Highly Crystalline Sodium Manganese Ferrocyanide Microcubes for Advanced Sodium Ion Battery Cathodes. *J. Mater. Chem. A Mater.* **2019**, *7*, 22248–22256. [CrossRef]
31. Duan, C.; Meng, Y.; Wang, Y.; Zhang, Z.; Ge, Y.; Li, X.; Guo, Y.; Xiao, D. High-Crystallinity and High-Rate Prussian Blue Analogues Synthesized at the Oil–Water Interface. *Inorg. Chem. Front.* **2021**, *8*, 2008–2016. [CrossRef]
32. Wang, B.; Han, Y.; Wang, X.; Bahlawane, N.; Pan, H.; Yan, M.; Jiang, Y. Prussian Blue Analogs for Rechargeable Batteries. *iScience* **2018**, *3*, 110–133. [CrossRef]
33. Jung, S.M.; Lim, Y.J.; Kwon, J.; Wei, K.; Lee, J.; Kim, K.S.; Kang, S.G.; Kim, Y.T. Low-Hysteresis Manganese Hexacyanoferrate (MnHCF) Aqueous Battery for Low-Grade Thermal Energy Harvesting. *J. Power Sources* **2022**, *524*, 231080. [CrossRef]
34. Ju, H.; Lang, H.; Yi, T.; Tian, K.; Yue, J.; Hu, L.; Zhao, L.; Liu, S.; Kong, D. Solid-State Synthesis of Diminutively Granular and Highly Crystalline Manganese Hexacyanoferrate for a Supercapacitor Electrode. *CrystEngComm* **2024**, *26*, 1133–1140. [CrossRef]
35. Lu, Y.; Wang, L.; Cheng, J.; Goodenough, J.B.; Zhou, F.C.; Sun, Y.H.; Li, J.Q.; Nan, J.M.; Wang, X.J.; Krumeich, F.; et al. Prussian Blue: A New Framework of Electrode Materials for Sodium Batteries. *Chem. Commun.* **2012**, *48*, 6544–6546. [CrossRef] [PubMed]
36. Wang, X.J.; Krumeich, F.; Nesper, R. Nanocomposite of Manganese Ferrocyanide and Graphene: A Promising Cathode Material for Rechargeable Lithium Ion Batteries. *Electrochem. Commun* **2013**, *34*, 246–249. [CrossRef]
37. Zhou, F.C.; Sun, Y.H.; Li, J.Q.; Nan, J.M. K 1−x Mn 1+x/2 [Fe(CN) 6]·yH 2 O Prussian Blue Analogues as an Anode Material for Lithium-Ion Batteries. *Appl. Surf. Sci.* **2018**, *444*, 650–660. [CrossRef]
38. Heo, J.W.; Chae, M.S.; Hyoung, J.; Hong, S.T. Rhombohedral Potassium-Zinc Hexacyanoferrate as a Cathode Material for Nonaqueous Potassium-Ion Batteries. *Inorg. Chem.* **2019**, *58*, 3065–3072. [CrossRef] [PubMed]
39. Zhang, L.; Chen, L.; Zhou, X.; Liu, Z. Towards High-Voltage Aqueous Metal-Ion Batteries beyond 1.5 V: The Zinc/Zinc Hexacyanoferrate System. *Adv. Energy Mater.* **2015**, *5*, 1400930. [CrossRef]
40. Zhang, L.; Chen, L.; Zhou, X.; Liu, Z. Morphology-Dependent Electrochemical Performance of Zinc Hexacyanoferrate Cathode for Zinc-Ion Battery. *Sci. Rep.* **2015**, *5*, 18263. [CrossRef]
41. Lee, S.H.; Huh, Y.D. Preferential Evolution of Prussian Blue's Morphology from Cube to Hexapod. *Bull. Korean Chem. Soc.* **2012**, *33*, 1078–1080. [CrossRef]
42. Coates, J. Interpretation of Infrared Spectra, A Practical Approach. In *Encyclopedia of Analytical Chemistry*; John Wiley & Sons Ltd.: Chichester, UK, 2000; ISBN 9780470027318.

43. Lejeune, J.; Brubach, J.B.; Roy, P.; Bleuzen, A. Application of the Infrared Spectroscopy to the Structural Study of Prussian Blue Analogues. *Comptes Rendus Chim.* **2014**, *17*, 534–540. [CrossRef]
44. Fang, Z.; Yin, Y.; Qiu, X.; Zhu, L.; Dong, X.; Wang, Y.; Xia, Y. Prussian Blue Cathode with Intercalation Pseudocapacitive Behavior for Low-Temperature Batteries. *Adv. Energy Sustain. Res.* **2021**, *2*, 2100105. [CrossRef]
45. Gerber, S.J.; Erasmus, E. Electronic Effects of Metal Hexacyanoferrates: An XPS and FTIR Study. *Mater. Chem. Phys.* **2018**, *203*, 73–81. [CrossRef]
46. Ojwang, D.O.; Grins, J.; Wardecki, D.; Valvo, M.; Renman, V.; Häggström, L.; Ericsson, T.; Gustafsson, T.; Mahmoud, A.; Hermann, R.P.; et al. Structure Characterization and Properties of K-Containing Copper Hexacyanoferrate. *Inorg. Chem.* **2016**, *55*, 5924–5934. [CrossRef] [PubMed]
47. Sato, O.; Einaga, Y.; Fujishima, A.; Hashimoto, K. Photoinduced Long-Range Magnetic Ordering of a Cobalt—Iron Cyanide. *Inorg. Chem.* **1999**, *38*, 4405–4412. [CrossRef]
48. Wang, X.; Wang, B.; Tang, Y.; Xu, B.B.; Liang, C.; Yan, M.; Jiang, Y. Manganese Hexacyanoferrate Reinforced by PEDOT Coating towards High-Rate and Long-Life Sodium-Ion Battery Cathode. *J. Mater. Chem. A Mater.* **2020**, *8*, 3222–3227. [CrossRef]
49. Mandal, M.; Chattopadhyay, K.; Bhattacharya, S.K. Sunlight Mediated Photocatalytic Reduction of Aqueous Cr(VI) Using Metal Hexacyanoferrate (M = Mn, Ni, Cu and Zn). *IOP Conf. Ser. Mater. Sci. Eng.* **2021**, *1080*, 012047. [CrossRef]
50. Deng, L.; Qu, J.; Niu, X.; Liu, J.; Zhang, J.; Hong, Y.; Feng, M.; Wang, J.; Hu, M.; Zeng, L.; et al. Defect-Free Potassium Manganese Hexacyanoferrate Cathode Material for High-Performance Potassium-Ion Batteries. *Nat. Commun.* **2021**, *12*, 2167. [CrossRef] [PubMed]
51. Zhang, J.; Deng, L.; Feng, M.; Zeng, L.; Hu, M.; Zhu, Y. Low-Defect K2Mn[Fe(CN)6]-Reduced Graphene Oxide Composite for High-Performance Potassium-Ion Batteries. *Chem. Commun.* **2021**, *57*, 8632–8635. [CrossRef]
52. Rodríguez-Hernández, J.; Reguera, E.; Lima, E.; Balmaseda, J.; Martínez-García, R.; Yee-Madeira, H. An Atypical Coordination in Hexacyanometallates: Structure and Properties of Hexagonal Zinc Phases. *J. Phys. Chem. Solids* **2007**, *68*, 1630–1642. [CrossRef]
53. Balmaseda, J.; Reguera, E.; Rodríguez-Hernández, J.; Reguera, L.; Autie, M. Behavior of Transition Metals Ferricyanides as Microporous Materials. *Microporous Mesoporous Mater.* **2006**, *96*, 222–236. [CrossRef]
54. Silva, M.N.T.; Ardisson, J.D.; Fabrisa, J.D.; Nossol, E. Zinc Hexacyanoferrate/Multi-Walled Carbon Nanotubes Films for Rechargeable Aqueous Batteries. *J. Braz. Chem. Soc.* **2020**, *31*, 1787–1795. [CrossRef]
55. Sciacca, R.; Zamponi, S.; Berrettoni, M.; Giorgetti, M. Stable Films of Zinc-Hexacyanoferrate: Electrochemistry and Ion Insertion Capabilities. *J. Solid. State Electrochem.* **2022**, *26*, 63–72. [CrossRef]
56. Duan, J.; Min, L.; Yang, T.; Chen, M.; Wang, C. High-Performance and Dendrite-Free Zn/KZnHCF Batteries Boosted by Aqueous Electrolyte Mixed with Ethanol. *J. Alloys Compd.* **2022**, *918*, 165619. [CrossRef]
57. González-Meza, O.A.; Larios-Durán, E.R.; Gutiérrez-Becerra, A.; Casillas, N.; Escalante, J.I.; Bárcena-Soto, M. Development of a Randles-Ševčík-like Equation to Predict the Peak Current of Cyclic Voltammetry for Solid Metal Hexacyanoferrates. *J. Solid. State Electrochem.* **2019**, *23*, 3123–3133. [CrossRef]
58. Li, W.; Xu, C.; Zhang, X.; Xia, M.; Yang, Z.; Yan, H.; Yu, H.; Zhang, L.; Shu, W.; Shu, J. Sodium Manganese Hexacyanoferrate as Zn Ion Host toward Aqueous Energy Storage. *J. Electroanal. Chem.* **2021**, *881*, 114968. [CrossRef]
59. Liu, Y.; Liu, H.; Zhang, R.; Zhong, Y.; Wu, Z.; Wang, X.; Zhang, Z. Recent Progress of Manganese-Based Prussian Blue Analogue Cathode Materials for Sodium-Ion Batteries. *Ionics* **2024**, *30*, 39–59. [CrossRef]
60. Li, M.; Sciacca, R.; Maisuradze, M.; Aquilanti, G.; Plaisier, J.; Berrettoni, M.; Giorgetti, M. Electrochemical Performance of Manganese Hexacyanoferrate Cathode Material in Aqueous Zn-Ion Battery. *Electrochim. Acta* **2021**, *400*, 139414. [CrossRef]
61. Maisuradze, M.; Li, M.; Gaboardi, M.; Aquilanti, G.; Rikkert Plaisier, J.; Giorgetti, M. Structural Modification of Partially Ni-Substituted MnHCF Cathode Material for Aqueous Zn-Ion Batteries. *Adv. Mater. Sci. Technol.* **2024**, *6*, 0629615. [CrossRef]
62. Kim, J.; Yi, S.H.; Li, L.; Chun, S.E. Enhanced Stability and Rate Performance of Zinc-Doped Cobalt Hexacyanoferrate (CoZnHCF) by the Limited Crystal Growth and Reduced Distortion. *J. Energy Chem.* **2022**, *69*, 649–658. [CrossRef]
63. Trócoli, R.; La Mantia, F. An Aqueous Zinc-Ion Battery Based on Copper Hexacyanoferrate. *ChemSusChem* **2015**, *8*, 481–485. [CrossRef] [PubMed]
64. Deng, W.; Li, Z.; Ye, Y.; Zhou, Z.; Li, Y.; Zhang, M.; Yuan, X.; Hu, J.; Zhao, W.; Huang, Z.; et al. Zn^{2+} Induced Phase Transformation of K2MnFe(CN)6 Boosts Highly Stable Zinc-Ion Storage. *Adv. Energy Mater.* **2021**, *11*, 2003639. [CrossRef]
65. Tan, Y.; Yang, H.; Miao, C.; Zhang, Y.; Chen, D.; Li, G.; Han, W. Hydroxylation Strategy Unlocking Multi-Redox Reaction of Manganese Hexacyanoferrate for Aqueous Zinc-Ion Battery. *Chem. Eng. J.* **2023**, *457*, 141323. [CrossRef]
66. Ni, G.; Xu, X.; Hao, Z.; Wang, W.; Li, C.; Yang, Y.; Zhou, C.; Qin, L.; Chen, W.; Yao, X.; et al. Tuning the Electrochemical Stability of Zinc Hexacyanoferrate through Manganese Substitution for Aqueous Zinc-Ion Batteries. *ACS Appl. Energy Mater.* **2021**, *4*, 602–610. [CrossRef]
67. Li, G.; Sun, L.; Zhang, S.; Zhang, C.; Jin, H.; Davey, K.; Liang, G.; Liu, S.; Mao, J.; Guo, Z. Developing Cathode Materials for Aqueous Zinc Ion Batteries: Challenges and Practical Prospects. *Adv. Funct. Mater.* **2024**, *34*, 2301291. [CrossRef]

nanomaterials

MDPI

Article

Ultra-Low Thermal Conductivity and Improved Thermoelectric Performance in Tungsten-Doped GeTe

Zhengtang Cai [1,2], Kaipeng Zheng [1,2], Chun Ma [1,2], Yu Fang [1,2], Yuyang Ma [1], Qinglin Deng [1,2,*] and Han Li [1,2,*]

1 School of Physics and Materials Science, Guangzhou University, Guangzhou 510006, China; zt.caine@e.gzhu.edu.cn (Z.C.); 2112119106@e.gzhu.edu.cn (K.Z.); 2112119081@e.gzhu.edu.cn (C.M.); 2112219110@e.gzhu.edu.cn (Y.F.); 32119600037@e.gzhu.edu.cn (Y.M.)

2 Research Center for Advanced Information Materials (CAIM), Huangpu Research & Graduate School of Guangzhou University, Guangzhou 510555, China

* Correspondence: qldeng@gzhu.edu.cn (Q.D.); lihan@gzhu.edu.cn (H.L.)

Abstract: Compared to SnTe and PbTe base materials, the GeTe matrix exhibits a relatively high Seebeck coefficient and power factor but has garnered significant attention due to its poor thermal transport performance and environmental characteristics. As a typical p-type IV–VI group thermoelectric material, W-doped GeTe material can bring additional enhancement to thermoelectric performance. In this study, the introduction of W, $Ge_{1-x}W_xTe$ (x = 0, 0.002, 0.005, 0.007, 0.01, 0.03) resulted in the presence of high-valence state atoms, providing additional charge carriers, thereby elevating the material's power factor to a maximum PF_{peak} of approximately 43 μW cm^{-1} K^{-2}, while slightly optimizing the Seebeck coefficient of the solid solution. Moreover, W doping can induce defects and promote slight rhombohedral distortion in the crystal structure of GeTe, further reducing the lattice thermal conductivity κ_{lat} to as low as approximately 0.14 W m^{-1} K^{-1} (x = 0.002 at 673 K), optimizing it to approximately 85% compared to the GeTe matrix. This led to the formation of a p-type multicomponent composite thermoelectric material with ultra-low thermal conductivity. Ultimately, W doping achieves the comprehensive enhancement of the thermoelectric performance of GeTe base materials, with the peak ZT value of sample $Ge_{0.995}W_{0.005}Te$ reaching approximately 0.99 at 673 K, and the average ZT optimized to 0.76 in the high-temperature range of 573–723 K, representing an increase of approximately 17% compared to pristine GeTe within the same temperature range.

Keywords: GeTe; ultra-low thermal conductivity; synergistic effects; multiscale materials; electronic engineering

1. Introduction

Thermoelectric (TE) materials have attracted considerable attention as environmentally friendly and green energy materials due to their ability to directly convert thermal energy into electrical energy. The performance of TE materials is typically assessed using the dimensionless figure of merit ZT, defined as ZT = $S^2\sigma T/\kappa_{tot}$, where S represents the Seebeck coefficient, σ is the electrical conductivity, T is the temperature in Kelvin, and κ_{tot} denotes the total thermal conductivity (the sum of the lattice thermal conductivity and electronic thermal conductivity, i.e., $\kappa_{tot} = \kappa_{lat} + \kappa_e$). The optimization of TE material performance primarily involves tuning the carrier concentration to enhance the electrical transport properties and improving the phonon transport characteristics to reduce the lattice thermal conductivity [1–3], thereby increasing the power factor and ultimately enhancing the thermoelectric performance of the material [4,5].

In recent years, GeTe has emerged as a typical IV–VI group semiconductor thermoelectric material [6]. Benefiting from its crystal structure closely resembling that of SnTe [7–9] and PbTe [10–12], GeTe exhibits a relatively high intrinsic Seebeck coefficient. However, due to the environmental toxicity of PbTe, poor mechanical and thermoelectric properties of SnTe, and the fact that the peak ZT value of GeTe-based materials can exceed 0.8 in the temperature range of 300–750 K, GeTe has become a hot topic in thermoelectric research. Nevertheless, the intrinsic electrical properties of GeTe are generally suboptimal, its thermal stability is inadequate, and its thermal transport performance is poor, limiting its widespread application. GeTe possesses a narrow bandgap structure and exhibits different crystal structures at different temperatures: it adopts a rhombohedral crystal structure (r-GeTe) at room temperature and a cubic crystal structure (c-GeTe) at high temperatures, with a phase transition from r- to c-GeTe occurring around ~700 K [3,13]. We also realized that a slight rhombohedral distortion occurs along the [1 1 1] crystal axis direction [14], which results in the convergence of the valence band [8,12,15]. Nevertheless, the phase transition is detrimental to the thermoelectric properties of GeTe materials as it likely results in the additional loss of electron energy due to extensive lattice deformation. This, in turn, negatively impacts carrier mobility, affecting the figure of merit (ZT). Essentially, the primary goal of thermoelectric material optimization is to increase the ZT value, which can be achieved by reasonably regulating the coupling relationship between the three thermoelectric performance parameters: S, σ, and κ_{tot}. When the electrical conductivity of the sample is increased, its electronic thermal conductivity also increases. However, excessively high electrical conductivity has a detrimental impact on carrier mobility, resulting in a decreasing Seebeck coefficient and affecting the thermoelectric properties of the material. Therefore, the key to improving the ZT value of materials lies in synergistically regulating the electrical and thermal transport properties. In order to optimize the thermoelectric properties of GeTe-based materials, researchers typically employ doping or alloying techniques. In doping engineering, the carrier concentration and energy band structure of the material can be optimized by applying donor doping, which results in a larger Seebeck coefficient and a higher power factor. Furthermore, doping engineering can result in the introduction of point defects, dislocations, and grain boundary defects, which lead to lattice distortions in GeTe material. This intensifies phonon scattering within the material and reduces the lattice's thermal conductivity, thereby further increasing the ZT value.

In our previous work, it was demonstrated that tungsten (W) doping can effectively enhance the electrical conductivity of the material, resulting in the optimization of a higher power factor and the capacity to elevate the Seebeck coefficient [16]. Regarding thermal transport properties, previous research has demonstrated that W doping can also introduce a multitude of grain boundary and dislocation defects, a considerable number of which can enhance phonon scattering in the material, resulting in a significant reduction in the lattice thermal conductivity and, ultimately, a dramatic optimization of the ZT value.

Consequently, the present study aims to achieve the same optimization via the elemental tungsten (W) doping of a GeTe system [17]. It is noteworthy that the electrical and thermal transport properties of GeTe-based materials can be modulated by designing W substitution (donor doping). Firstly, the intervention of W substitution introduced high-valence atoms, resulting in an increase in the power factor of the samples across the entire temperature range due to the introduction of additional carriers. Secondly, W also exists in the solid solution of $Ge_{1-x}W_xTe$ (x = 0, 0.002, 0.005, 0.007, 0.01, 0.03) in the form of W atoms [18–20], forming defects to replace some of the Ge vacancies [21–23]. On the other hand, the formation of defects further strengthens the scattering of phonons in the material [24–26], significantly reducing the lattice thermal conductivity. Ultimately, we achieved a significant improvement in the peak power factor (PF_{peak}) at 673 K up to ~43 $\mu W\ cm^{-1}\ K^{-2}$ and the optimization of the ultra-low lattice thermal conductivity of 0.14 $W\ m^{-1}\ K^{-1}$ at 673 K for a W doping concentration of x = 0.002. This contributes to a remarkable enhancement in the average ZT value up to ~0.76 within the temperature range of 573–723 K and represents a 17% increase compared to the GeTe matrix, thereby improving the thermoelectric performance of GeTe-based materials overall [7,9,27].

2. Materials and Methods

2.1. Materials and Synthesis

This study involved the preparation of $Ge_{1-x}W_xTe$ (x = 0, 0.002, 0.005, 0.007, 0.01, 0.03) compounds in an inert gas (Ar) atmosphere within a glovebox. High-purity elements Ge (99.999%, Aladdin, Shanghai, China), W (99.99%, Aladdin), and Te (99.99%, Aladdin) were accurately weighed and transferred into graphite crucibles, thoroughly mixed, and then packed into quartz glass tubes. Subsequently, the tubes were sealed under high vacuum conditions (below 10^{-5} Torr). The loaded quartz tubes were then placed in a box furnace and slowly heated to 1373 K, maintained for 12 h, followed by quenching with water to room temperature. Subsequently, the temperature was raised again to 923 K for annealing over 3 days, and finally slowly cooled to room temperature. The resulting alloy ingots were carefully removed and manually ground into fine powders using an agate mortar and pestle. Appropriate amounts of the ground powders were weighed and loaded into prepared graphite molds, which were then placed into a spark plasma sintering furnace (SPS, KCE-FCT HPD10, Sachsenheim, Germany) for vacuum hot-pressing sintering. During the sintering process, the samples were subjected to a uniaxial pressure of 60 MPa in a high vacuum, heated to 573 K in 5 min, and then heated to 823 K in 3 min. Finally, they were held for 5 min before being removed and allowed to cool naturally.

2.2. Measurement and Characterizations

After SPS sintering and subsequent cooling, the samples were subjected to wire cutting to produce rectangular prismatic specimens measuring 12 × 3 × 3 mm. These specimens were then lightly polished and placed into a four-probe thermoelectric measurement system (CTA, CRYOALL CTA-3S, Beijing, China) to simultaneously measure the temperature-dependent electrical conductivity σ and Seebeck coefficient S. Additionally, the thin discs obtained from SPS sintering underwent thermal diffusivity testing using the Laser-Flash Thermal Conductivity Instrument (LFA 467 HT NETZSCH, Tannesstein, Germany) to determine the thermal diffusivity coefficient D of the samples. Subsequently, the density ρ of each sample was determined using the Archimedes' displacement method. The total thermal conductivity κ_{tot} of the samples was calculated using the formula $\kappa_{tot} = D\rho C_p$, where C_p represents the theoretical specific heat capacity of the GeTe material. Furthermore, the electronic thermal conductivity κ_e of the samples was calculated according to the Wiedemann–Franz law and the formula $\kappa_e = L\sigma T$. Finally, the lattice thermal conductivity κ_{lat} of the samples was obtained by subtracting κ_e from κ_{tot}. Subsequently, X-ray diffraction (XRD) analysis was conducted at room temperature using a Rigaku Smartlab 9 KW (Tokyo, Japan) instrument to perform the compositional analysis of the samples. To further understand the influence of the microstructural composition of the samples on the thermoelectric properties of the GeTe materials, a transmission electron microscopy (TEM, FEI HELIOS 5CX, Hillsboro, OR, USA) characterization test was performed to observe and analyze the mechanism.

3. Results and Discussion

The room temperature powder X-ray diffraction (XRD) results of the $Ge_{1-x}W_xTe$ samples (x = 0, 0.002, 0.005, 0.007, 0.01, 0.03) are depicted in Figure 1a. It can be observed that upon W doping into GeTe, most of the diffraction peaks match well with the rhombohedral phase single-phase structure of r-GeTe (space group: $R3m$, PDF: 47-1079). In Figure 2a,b, the slight precipitation of Ge can be observed at low angles of 25~26° and 42~44°, respectively. This phenomenon is commonly observed throughout the entire GeTe system and is attributed to the presence of large Ge vacancies in the GeTe system, consistent with previous reports [3,28,29], which do not significantly affect the ultimate thermoelectric performance of GeTe-based materials [2,24,30]. The appearance of dual peaks of GeTe at this angle further confirms its rhombohedral structure at room temperature. Moreover, the sample $Ge_{0.97}W_{0.03}Te$ exhibits a distinct second-phase diffraction peak at ~40° in Figure 1b. Through XRD spectrum analysis and data comparison, the main component of the diffraction peak at this angle is identified as W (space group: *Im*3*m*, PDF: 04-0806), indicating a solubility of W in the GeTe system of approximately 0.5 mol%. Additionally, GeTe undergoes a phase transition from the r- to the c-GeTe (c-GeTe, space group: *Fm*-3*m*) at ~700 K, transforming its crystal structure along the [1 1 1] direction into a cubic structure, which is similar to that of PbTe [31] and SnTe [1,20]. This cubic structure is also similar to the crystal structure of the precipitated W phase in this study, suggesting structural similarity with GeTe [32].

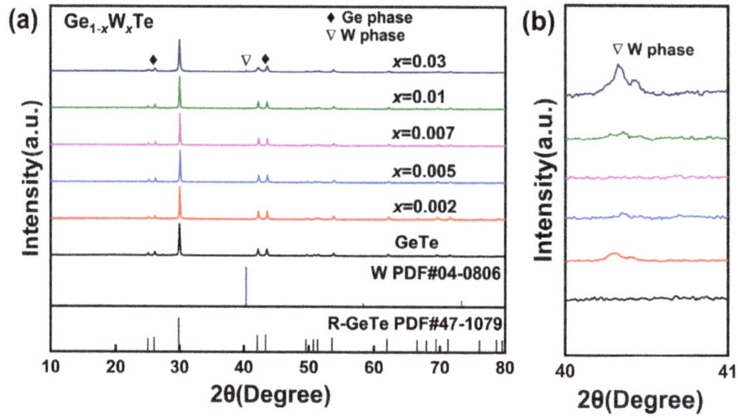

Figure 1. (**a**) Room temperature powder XRD patterns of the $Ge_{1-x}W_xTe$ (x = 0, 0.002, 0.005, 0.007, 0.01, 0.03) samples; (**b**) enlarged view of the W phase at 40~41°.

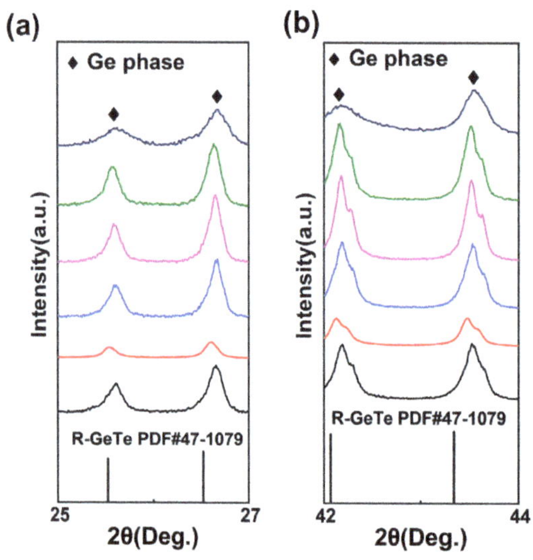

Figure 2. Room temperature powder XRD patterns enlarged view of the Ge phase of the $Ge_{1-x}W_xTe$ (x = 0, 0.002, 0.005, 0.007, 0.01, 0.03) samples (**a**) at 25~27°; (**b**) enlarged view at 42~44°.

Upon W doping in GeTe, the electrical transport properties of the material undergo subtle changes. In Figure 3a, as the proportion of W doping increases, the electrical conductivity (σ) exhibits a trend of initially decreasing and then increasing. Compared to the GeTe matrix, when x = 0.002, the σ of the sample $Ge_{0.998}W_{0.002}Te$ decreases to ~1.8×10^3 S cm^{-1} at 723 K. When $x > 0.005$, the σ of the sample $Ge_{1-x}W_xTe$ increases with the amount of W doping, and at x = 0.03, the σ of the sample $Ge_{0.97}W_{0.03}Te$ is higher than that of other samples and higher than the GeTe matrix overall. This may be attributed to the introduction of the high-valence state W into GeTe, providing additional charge carriers and resulting in an increase in the carrier concentration. In Figure 3c, a schematic diagram of the Seebeck coefficient (S) of the sample $Ge_{1-x}W_xTe$ (x = 0, 0.002, 0.005, 0.007, 0.01, 0.03) as a function of temperature is presented. In the mid-low temperature range (300–573 K), the S of the sample $Ge_{0.995}W_{0.005}Te$ shows a certain improvement, reaching ~154 μV K^{-1} at 723 K. When $x \geq 0.005$, S begins to decline, especially when x = 0.03, the sample $Ge_{0.97}W_{0.03}Te$ exhibits a significant reduction in S in the high-temperature range (600–750 K). This may be because W has reached its solubility limit in the GeTe system, existing in the material structure in the form of W. Additionally, as W belongs to heavy atoms, it affects carrier mobility. In Figure 3d, the power factor PF of the sample $Ge_{0.998}W_{0.002}Te$ is lower than that of the undoped sample GeTe, and when the doping level rises to x = 0.005, the PF of the sample $Ge_{0.995}W_{0.005}Te$ is higher than that of the undoped GeTe matrix, with a peak value reaching ~43 μW cm^{-1} K^{-2}. When the W doping ratio is greater than x = 0.005, the PF of the sample $Ge_{1-x}W_xTe$ decreases with the increase in the doping concentration. This phenomenon occurs because W doping introduces additional heavy atoms into the GeTe material, significantly affecting the electrical transport properties of the sample [5,33,34]. Despite the weak increase in the Seebeck coefficient of the sample $Ge_{0.995}W_{0.005}Te$, W doping leads to a higher peak PF in the material compared to the GeTe matrix.

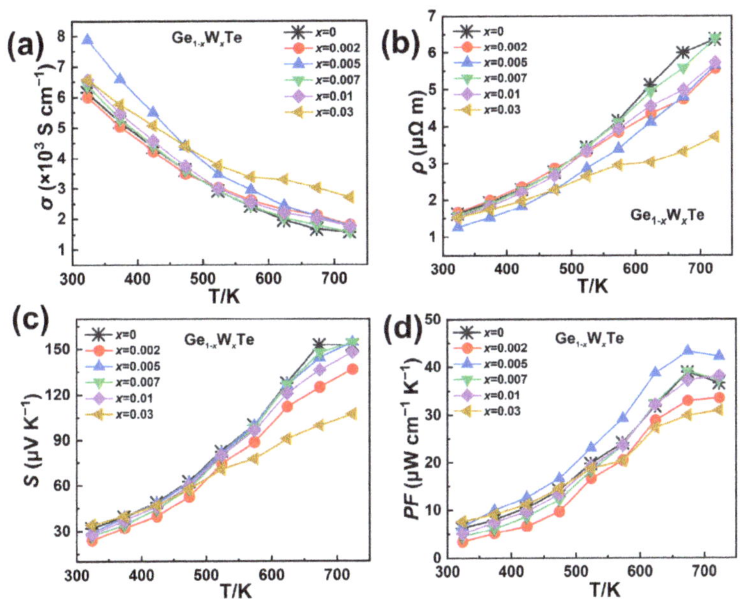

Figure 3. Temperature-dependent electrical performance of the $Ge_{1-x}W_xTe$ (x = 0, 0.002, 0.005, 0.007, 0.01, 0.03) solid solution; (**a**) electrical conductivity σ; (**b**) resistivity ρ; (**c**) Seebeck coefficient S; (**d**) power factor PF.

In Figure 4a, for the $Ge_{1-x}W_xTe$ samples (x = 0, 0.002, 0.005, 0.007, 0.01, 0.03), at x = 0.002, the total thermal conductivity κ_{tot} exhibits a slight decrease compared to the undoped sample, while at x = 0.005, the change in κ_{tot} of the sample $Ge_{0.995}W_{0.005}Te$ is minimal, differing only marginally from the sample at x = 0.002. As the doping ratio continues to increase, the overall trend of κ_{tot} continues to decrease. At x = 0.03, the κ_{tot} of the sample $Ge_{0.97}W_{0.03}Te$ remains relatively flat, and thermal performance deteriorates. This phenomenon is consistent with the earlier XRD findings, where the degradation is attributed to the different forms of W presence in the GeTe system, which significantly affects the carriers, ultimately influencing the thermal transport properties. Additionally, due to the higher valence state of W compared to Ge^{2+}, it may introduce additional carriers into the system, further increasing the carrier concentration of the GeTe base material, resulting in a higher electronic thermal conductivity κ_e compared to undoped GeTe, with higher proportions of W leading to higher κ_e and a flattening trend. In Figure 4b, the slope of the κ_e curve for the sample $Ge_{0.97}W_{0.03}Te$ decreases, becoming flatter overall. When the W doping ratio reaches x = 0.005, the κ_e of the sample $Ge_{0.995}W_{0.005}Te$ reaches ~5.8 W m^{-1} K^{-1} at temperatures near room temperature to 323 K, 29% higher than other samples with different W doping compositions. In Figure 4c, due to the generally higher electronic thermal conductivity κ_e of the $Ge_{1-x}W_xTe$ samples compared to the undoped GeTe base material, the lattice thermal conductivity κ_{lat} of the samples decreases significantly. For instance, the κ_{lat} of the sample $Ge_{0.998}W_{0.002}Te$ reaches a minimum value of ~0.14 W m^{-1} K^{-1} at 673 K, while the κ_{lat} of the sample $Ge_{0.995}W_{0.005}Te$ decreases to ~0.44 W m^{-1} K^{-1} at 673 K. The effective reduction of κ_{lat} in the material is achieved via appropriate W doping into the GeTe system.In Figure 4d, the extremely low κ_{lat} of the sample $Ge_{0.998}W_{0.002}Te$ reduces by approximately 85%, 81%, 79%, 77%, 74%, and 39% compared to the works of Yang et al. on GeTe: Bi, In [35], Liu et al. on GeTe: Mn, Bi [13], Li et al. on GeTe: Cd, Bi [36], Srinivasan et al. on GeTe: Ag [37], Xu et al. on GeTe: Se, Bi [38], and Shuai et al. on GeTe: Ti, Bi [39], respectively.

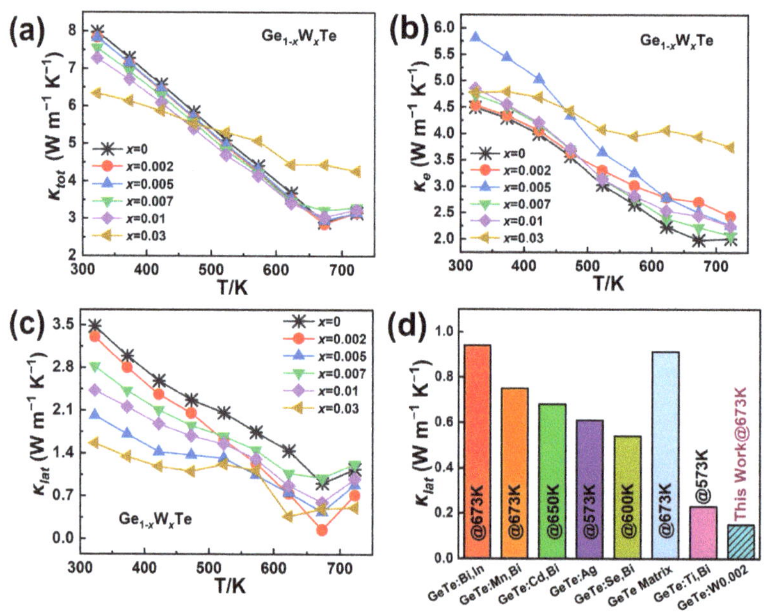

Figure 4. Thermal transport properties of the $Ge_{1-x}W_xTe$ (x = 0, 0.002, 0.005, 0.007, 0.01, 0.03) samples; (**a**) total thermal conductivity κ_{tot}; (**b**) electronic thermal conductivity κ_e; (**c**) lattice thermal conductivity κ_{lat}; (**d**) lattice thermal conductivity of this work compared to the GeTe matrix; GeTe: Bi, In [35]; GeTe: Mn, Bi [13]; GeTe: Cd, Bi [36]; GeTe: Ag [37]; GeTe: Se, Bi [38]; and GeTe: Ti, Bi [39].

In order to further investigate the mechanism of the influence of W doping on the thermoelectric transport properties of the GeTe system and the microstructural morphology of the material, we performed a transmission electron microscopy (TEM) characterization test on the $Ge_{0.995}W_{0.005}Te$ sample. As shown in Figure 5a, a low magnification TEM image of the sample $Ge_{0.995}W_{0.005}Te$ is demonstrated, and the typical herringbone fishbone structure of the GeTe material can be clearly observed [40], which is uniformly interlaced throughout the sample. The magnified image of the orange dashed box area in Figure 5a corresponds to that shown in Figure 5b, and the high-resolution image can better illustrate the herringbone fishbone structure, from which it can be seen that the W doping of GeTe does not produce a second phase to the microstructure of the material. It also improves the orderliness of the atomic arrangement of the GeTe material. Subsequently, selected area electron diffraction (SAED) was performed on the same sample by choosing the appropriate region. The SAED image of the $Ge_{0.995}W_{0.005}Te$ sample at high magnification along the direction of the [2 1 1] crystal axis is demonstrated in Figure 5c, from which the diffractograms show that the synthesized sample has a single-crystalline structure and is consistent with the r-GeTe crystalline phase. The angle between the $(\bar{1},1,1)$, $(\bar{2},2,2)$, and $(\bar{1},0,2)$ crystal planes is ~110.7°, which is slightly smaller than the 120° angle of the GeTe rhombic phase structure, which precisely indicates that the W doping is introduced to more likely substitute the Ge sites and bring about slight nanoscale lattice distortions [41], which agrees with the results of the sample's electrical conductivity test in Figure 3d.

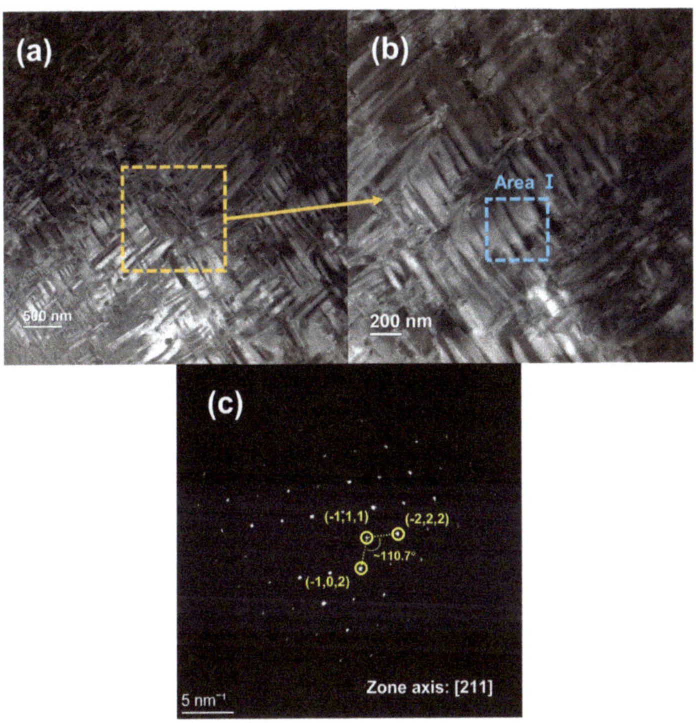

Figure 5. Transmission scanning microscopy (TEM) images of the $Ge_{0.995}W_{0.005}Te$ solid solution: (**a**) low magnification TEM image; (**b**) enlarged TEM image of the area surrounded by dashed lines in (**a**); (**c**) selected area electron diffraction (SAED) pattern of the GeTe matrix with the [2 1 1] zone axis.

The blue dashed box area in Figure 5b of the $Ge_{0.995}W_{0.005}Te$ sample was selected to be photographed at high magnification to obtain a high-resolution magnified image, as shown in Figure 6a. It clearly demonstrates that the $Ge_{0.995}W_{0.005}Te$ sample exhibits a pronounced dislocation structure in Figure 6a, resulting from the disparate mass and size of the Ge and W atoms. This leads to the typical lattice distortions observed in W doping at the Ge sites, which introduces defects, optimizes the electron–phonon transport properties of the material, and further enhances the phonon scattering. This is consistent with the findings of our previous work [16], which demonstrated that W is capable of introducing a significant number of defects, thereby greatly reducing the lattice thermal conductivity of the material. In the high-resolution image of Figure 6b, the area within the yellow dashed box represents the interplanar spacing measured for sample $Ge_{0.995}W_{0.005}Te$, as indicated by the yellow arrows corresponding to Figure 6c,d, which represent the d spacing values of $d_{(202)}$~0.3012 nm and $d_{(202)}$~0.3018 nm, respectively. They are both very close to the standard value of the d spacing of the r-GeTe (202) planar of 2.991 Å. This implies that W doping does not bring about a significant second phase but rather modulates the microstructure of the GeTe material by introducing defects. It is noteworthy that these results are in agreement with the previously mentioned XRD and electrical-thermal performance tests, which significantly enhance the electrical conductivity of the GeTe material.

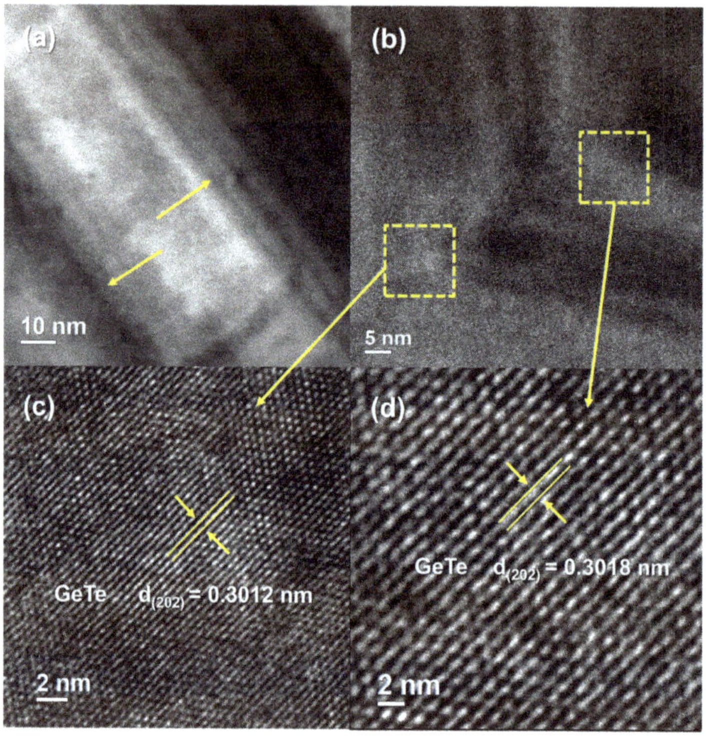

Figure 6. (**a**) High magnification TEM image of the blue dashed box area (area I) corresponding to Figure 5b of the $Ge_{0.995}W_{0.005}Te$ solid solution; (**b**) HRTEM image of the same solid solution; (**c**,**d**) enlarged HRTEM images corresponding to the area surrounded by yellow dashed lines in (**a**), respectively.

Figure 7 depicts the temperature-dependent curves of the sample $Ge_{1-x}W_xTe$ (x = 0, 0.002, 0.005, 0.007, 0.01, 0.03). When the W doping concentration reaches x = 0.005, the ZT value of the sample $Ge_{0.995}W_{0.005}Te$ reaches its maximum, approximately 0.99 at 673 K, 10% higher than the undoped sample. Due to the high melting point of W, in the high-temperature range (573–723 K), the average ZT value of the $Ge_{1-x}W_xTe$ sample is approximately 0.76, 17% higher than the undoped GeTe matrix at the same temperature range. However, a high dopant concentration ($x \geq 0.007$) causes an excess of W in the sample, which leads to a decrease in the carrier mobility due to increased carrier scattering; at the same time, the electronic thermal conductivity κ_e increases (as well as the total thermal conductivity κ_{tot}), leading to a negative optimization in the figure of merit. Therefore, the sample $Ge_{1-x}W_xTe$ achieves the highest ZT value for the x = 0.005 doping concentration.

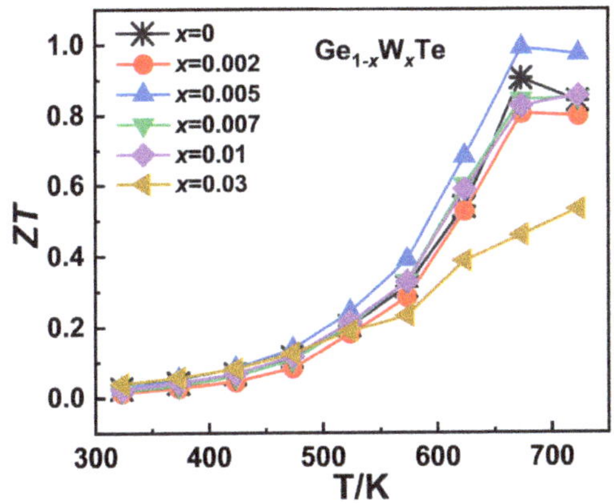

Figure 7. Temperature-dependent trend of the figure of merit (ZT) value of the $Ge_{1-x}W_xTe$ (x = 0, 0.002, 0.005, 0.007, 0.01, 0.03) samples.

4. Conclusions

In summary, this study elucidates the advantages and disadvantages of W-doped GeTe materials, analyzing the influence of W on the material structure, electrical transport properties, and thermal transport properties. The higher valence state of W allows for the introduction of additional charge carriers, thereby enhancing the electrical conductivity of the material while simultaneously increasing the power factor, thereby improving the Seebeck coefficient. With a higher melting point than GeTe, W stabilizes the rock-salt cubic structure of GeTe (achieved through the phase transition from r-GeTe) near 700 K. It is noteworthy that W doping can also substitute the Ge atoms in the GeTe material system, introducing dislocation defects to optimize the thermal transport performance. W doping can also induce lattice distortions in the GeTe structure [15], enhancing phonon scattering and effectively reducing the lattice thermal conductivity of the material at suitable W doping concentrations, resulting in GeTe-based materials with ultra-low lattice thermal conductivity as low as approximately 0.14 W m^{-1} K^{-1} at 673 K [16,27,42]. Compared to the previous work [35], this study demonstrates a maximum reduction of approximately 85% in lattice thermal conductivity. The experimental results confirmed that W doping not only introduces high-valence state W atoms to enhance the power factor of the sample but also simultaneously reduces the lattice thermal conductivity by optimizing the material structure [32], thereby comprehensively regulating the thermoelectric transport properties of GeTe-based materials. Ultimately, the average ZT value of the sample $Ge_{0.995}W_{0.005}Te$ reached ~0.76 in the temperature range of 573–723 K, approximately 17% higher than undoped GeTe. Furthermore, the peak ZT value of sample $Ge_{0.995}W_{0.005}Te$ reached ~0.99 across the entire temperature range, representing a ~10% improvement over the GeTe base material.

Author Contributions: Investigation, methodology, and data curation, Z.C. and K.Z.; Visualization and investigation, C.M., Y.M. and Y.F.; Writing—original draft preparation, Z.C.; Writing—review and editing, Z.C., Q.D. and H.L.; Conceptualization, H.L.; Supervision, H.L.; Funding acquisition, H.L. All authors have read and agreed to the published version of the manuscript.

Funding: This work is funded by the Young Talent Program of Guangzhou University (Grant No. RQ2020077), the Key Discipline of Materials Science and Engineering and Bureau of Education of Guangzhou Municipality (Grant No. 202201020144), Bureau of Education of Guangzhou Municipality (Grant No. 202255464) and the China Biosafety Research Special Program (Grant No. 20SWAQX40).

Data Availability Statement: Data are contained within the article.

Conflicts of Interest: The authors declare no conflicts of interest.

References

1. Mao, J.; Liu, Z.H.; Zhou, J.W.; Zhu, H.T.; Zhang, Q.; Chen, G.; Ren, Z.F. Advances in thermoelectrics. *Adv. Phys.* **2018**, *67*, 69–147. [CrossRef]
2. Yang, L.; Chen, Z.G.; Dargusch, M.S.; Zou, J. High Performance Thermoelectric Materials: Progress and Their Applications. *Adv. Energy Mater.* **2017**, *8*, 1701797–1701824. [CrossRef]
3. Zhang, X.; Bu, Z.; Lin, S.; Chen, Z.; Li, W.; Pei, Y. GeTe Thermoelectrics. *Joule* **2020**, *4*, 986–1003. [CrossRef]
4. Liu, Z.H.; Gao, W.H.; Zhang, W.H.; Sato, N.; Guo, Q.S.; Mori, T. High Power Factor and Enhanced Thermoelectric Performance in Sc and Bi Codoped GeTe: Insights into the Hidden Role of Rhombohedral Distortion Degree. *Adv. Energy Mater.* **2020**, *10*, 202002588–202002597. [CrossRef]
5. Hong, M.; Lyv, W.Y.; Li, M.; Xu, S.D.; Sun, Q.; Zou, J.; Chen, Z.G. Rashba Effect Maximizes Thermoelectric Performance of GeTe Derivatives. *Joule* **2020**, *4*, 2030–2043. [CrossRef]
6. Srinivasan, B.; Le Tonquesse, S.; Gellé, A.; Bourgès, C.; Monier, L.; Ohkubo, I.; Halet, J.F.; Berthebaud, D.; Mori, T. Screening of transition (Y, Zr, Hf, V, Nb, Mo, and Ru) and rare-earth (La and Pr) elements as potential effective dopants for thermoelectric GeTe—An experimental and theoretical appraisal. *J. Mater. Chem. A* **2020**, *8*, 19805–19821. [CrossRef]
7. Kihoi, S.K.; Shenoy, U.S.; Kahiu, J.N.; Kim, H.; Bhat, D.K.; Lee, H.S. Ultralow Lattice Thermal Conductivity and Enhanced Mechanical Properties of Cu and Sb Co-Doped SnTe Thermoelectric Material with a Complex Microstructure Evolution. *ACS Sustain. Chem. Eng.* **2022**, *10*, 1367–1372. [CrossRef]
8. Huo, H.; Wang, Y.; Xi, L.; Yang, J.; Zhang, W. The variation of intrinsic defects in XTe (X = Ge, Sn, and Pb) induced by the energy positions of valence band maxima. *J. Mater. Chem. C* **2021**, *9*, 5765–5770. [CrossRef]
9. Lei, K.; Huang, H.M.; Liu, X.J.; Wang, W.L.; Guo, K.; Zheng, R.K.; Li, H. Ultra-Low Lattice Thermal Conductivity Enables High Thermoelectric Properties in Cu and Y Codoped SnTe via Multi-Scale Composite Nanostructures. *ACS Sustain. Chem. Eng.* **2023**, *11*, 7541–7551. [CrossRef]
10. Ge, B.Z.; Lee, H.; Im, J.; Choi, Y.; Kim, S.Y.; Lee, J.Y.; Cho, S.P.; Sung, Y.E.; Choi, K.Y.; Zhou, C.J.; et al. Engineering an atomic-level crystal lattice and electronic band structure for an extraordinarily high average thermoelectric figure of merit in n-type PbSe. *Energy Environ. Sci.* **2023**, *16*, 3994–4008. [CrossRef]
11. Biswas, K.; He, J.; Zhang, Q.; Wang, G.; Uher, C.; Dravid, V.P.; Kanatzidis, M.G. Strained endotaxial nanostructures with high thermoelectric figure of merit. *Nat. Chem.* **2011**, *3*, 160–166. [CrossRef] [PubMed]
12. Pei, Y.; Shi, X.; LaLonde, A.; Wang, H.; Chen, L.; Snyder, G.J. Convergence of electronic bands for high performance bulk thermoelectrics. *Nature* **2011**, *473*, 66–69. [CrossRef] [PubMed]
13. Liu, Z.; Sun, J.; Mao, J.; Zhu, H.; Ren, W.; Zhou, J.; Wang, Z.; Singh, D.J.; Sui, J.; Chu, C.W.; et al. Phase-transition temperature suppression to achieve cubic GeTe and high thermoelectric performance by Bi and Mn codoping. *Proc. Natl. Acad. Sci. USA* **2018**, *115*, 5332–5337. [CrossRef] [PubMed]
14. Cahill, D.G.; Watson, S.K.; Pohl, R.O. Lower limit to the thermal conductivity of disordered crystals. *Phys. Rev. B Condens. Matter* **1992**, *46*, 6131–6140. [CrossRef] [PubMed]
15. Li, J.; Zhang, X.; Chen, Z.; Lin, S.; Li, W.; Shen, J.; Witting, I.T.; Faghaninia, A.; Chen, Y.; Jain, A.; et al. Low-Symmetry Rhombohedral GeTe Thermoelectrics. *Joule* **2018**, *2*, 976–987. [CrossRef]
16. Lei, K.; Feng, K.m.; Ma, C.; Cai, Z.T.; He, B.b.; Li, H. High power factor and ultra-low lattice thermal conductivity in $Sn_{1-x}W_xTe$ alloys via interstitial defects modulation. *J. Alloys Compd.* **2024**, *976*, 173187. [CrossRef]
17. Wang, G.; Zhou, J.; Sun, Z. First principles investigation on anomalous lattice shrinkage of W alloyed rock salt GeTe. *J. Phys. Chem. Solids* **2020**, *137*, 109220. [CrossRef]
18. Zheng, L.; Li, W.; Lin, S.; Li, J.; Chen, Z.; Pei, Y. Interstitial Defects Improving Thermoelectric SnTe in Addition to Band Convergence. *ACS Energy Lett.* **2017**, *2*, 563–568. [CrossRef]
19. Yin, L.C.; Liu, W.D.; Li, M.; Wang, D.Z.; Wu, H.; Wang, Y.F.; Zhang, L.X.; Shi, X.L.; Liu, Q.F.; Chen, Z.G. Interstitial Cu: An Effective Strategy for High Carrier Mobility and High Thermoelectric Performance in GeTe. *Adv. Funct. Mater.* **2023**, *33*, 2301750–2301758. [CrossRef]
20. Xin, J.; Li, S.; Yang, J.; Basit, A.; Long, Q.; Li, S.; Jiang, Q.; Xu, T.; Xiao, B. Tactfully decoupling interdependent electrical parameters via interstitial defects for SnTe thermoelectrics. *Nano Energy* **2020**, *67*, 104292. [CrossRef]
21. Foster, A.S.; Lopez Gejo, F.; Shluger, A.L.; Nieminen, R.M. Vacancy and interstitial defects in hafnia. *Phys. Rev. B* **2002**, *65*, 174117. [CrossRef]
22. Li, H.; Jing, H.; Han, Y.; Lu, G.-Q.; Xu, L.; Liu, T. Interfacial evolution behavior of $AgSbTe_{2.01}$/nanosilver/Cu thermoelectric joints. *Mater. Des.* **2016**, *89*, 604–610. [CrossRef]
23. Li, J.; Zhang, X.; Lin, S.; Chen, Z.; Pei, Y. Realizing the High Thermoelectric Performance of GeTe by Sb-Doping and Se-Alloying. *Chem. Mater.* **2016**, *29*, 605–611. [CrossRef]
24. Liu, H.; Zhang, X.; Li, J.; Bu, Z.; Meng, X.; Ang, R.; Li, W. Band and Phonon Engineering for Thermoelectric Enhancements of Rhombohedral GeTe. *ACS Appl. Mater. Interfaces* **2019**, *11*, 30756–30762. [CrossRef] [PubMed]
25. Su, L.; Wang, D.; Wang, S.; Qin, B.; Wang, Y.; Qin, Y.; Jin, Y.; Chang, C.; Zhao, L.D. High thermoelectric performance realized through manipulating layered phonon-electron decoupling. *Science* **2022**, *375*, 1385–1389. [CrossRef] [PubMed]

26. Li, M.R.; Ying, P.Z.; Du, Z.L.; Liu, X.L.; Li, X.; Fang, T.; Cui, J.L. Improved Thermoelectric Performance of P-type SnTe through Synergistic Engineering of Electronic and Phonon Transports. *ACS Appl. Mater. Interfaces* **2022**, *14*, 8171–8178. [CrossRef] [PubMed]
27. Xing, T.; Zhu, C.; Song, Q.; Huang, H.; Xiao, J.; Ren, D.; Shi, M.; Qiu, P.; Shi, X.; Xu, F.; et al. Ultralow Lattice Thermal Conductivity and Superhigh Thermoelectric Figure-of-Merit in (Mg, Bi) Co-Doped GeTe. *Adv. Mater.* **2021**, *33*, e2008773. [CrossRef] [PubMed]
28. Wu, Z.; Chen, X.; Mu, E.; Liu, Y.; Che, Z.; Dun, C.; Sun, F.; Wang, X.; Zhang, Y.; Hu, Z. Lattice Strain Enhances Thermoelectric Properties in Sb_2Te_3/Te Heterostructure. *Adv. Electron. Mater.* **2019**, *6*, 1900735. [CrossRef]
29. Wu, G.J.; Guo, Z.; Wang, R.Y.; Tan, X.J.; Cui, C.; Sun, P.; Hu, H.Y.; Wu, J.H.; Liu, G.Q.; Jiang, J. Structural modulation and resonant level enable high thermoelectric performance of GeTe in the mid-to-low temperature range. *J. Mater. Chem. A* **2023**, *11*, 20497–20505. [CrossRef]
30. Jin, Y.; Wang, D.Y.; Qiu, Y.T.; Zhao, L.D. Boosting the thermoelectric performance of GeTe by manipulating the phase transition temperature Sb doping. *J. Mater. Chem. C* **2021**, *9*, 6484–6490. [CrossRef]
31. Pei, Q.-X.; Guo, J.-Y.; Suwardi, A.; Zhang, G. Insights into interfacial thermal conductance in Bi_2Te_3-based systems for thermoelectrics. *Mater. Today Phys.* **2023**, *30*, 100953. [CrossRef]
32. Perumal, S.; Samanta, M.; Ghosh, T.; Shenoy, U.S.; Bohra, A.K.; Bhattacharya, S.; Singh, A.; Waghmare, U.V.; Biswas, K. Realization of High Thermoelectric Figure of Merit in GeTe by Complementary Co-doping of Bi and In. *Joule* **2019**, *3*, 2565–2580. [CrossRef]
33. Hong, M.; Zou, J.; Chen, Z.G. Thermoelectric GeTe with Diverse Degrees of Freedom Having Secured Superhigh Performance. *Adv. Mater.* **2019**, *31*, e1807071. [CrossRef] [PubMed]
34. Gao, W.H.; Liu, Z.H.; Zhang, W.H.; Sato, N.; Guo, Q.S.; Mori, T. Improved thermoelectric performance of GeTe via efficient yttrium doping. *Appl. Phys. Lett.* **2021**, *118*, 033901. [CrossRef]
35. Jin, Y.; Wang, D.; Zhu, Y.; Su, L.; Hong, T.; Wang, Z.; Ge, Z.-H.; Qiu, Y.; Zhao, L.-D. Contrasting roles of trivalent dopants M (M = In, Sb, Bi) in enhancing the thermoelectric performance of Ge0.94M0.06Te. *Acta Mater.* **2023**, *252*, 118926. [CrossRef]
36. Li, J.; Li, W.; Bu, Z.; Wang, X.; Gao, B.; Xiong, F.; Chen, Y.; Pei, Y. Thermoelectric Transport Properties of $Cd_xBi_yGe_{1-x-y}Te$ Alloys. *ACS Appl. Mater. Interfaces* **2018**, *10*, 39904–39911. [CrossRef] [PubMed]
37. Srinivasan, B.; Gautier, R.; Gucci, F.; Fontaine, B.; Halet, J.-F.; Cheviré, F.; Boussard-Pledel, C.; Reece, M.J.; Bureau, B. Impact of Coinage Metal Insertion on the Thermoelectric Properties of GeTe Solid-State Solutions. *J. Phys. Chem. C* **2017**, *122*, 227–235. [CrossRef]
38. Xu, L.; Wu, G.; Wang, R.; Yan, Z.; Cai, J.; Yang, J.; Wang, X.; Luo, J.; Tan, X.; Liu, G.; et al. Synergistically Optimized Thermal Conductivity and Carrier Concentration in GeTe by Bi-Se Codoping. *ACS Appl. Mater. Interfaces* **2022**, *14*, 14359–14366. [CrossRef] [PubMed]
39. Shuai, J.; Tan, X.J.; Guo, Q.; Xu, J.T.; Gellé, A.; Gautier, R.; Halet, J.F.; Failamani, F.; Jiang, J.; Mori, T. Enhanced thermoelectric performance through crystal field engineering in transition metal–doped GeTe. *Mater. Today Phys.* **2019**, *9*, 100094. [CrossRef]
40. Rinaldi, C.; Rojas-Sánchez, J.C.; Wang, R.N.; Fu, Y.; Oyarzun, S.; Vila, L.; Bertoli, S.; Asa, M.; Baldrati, L.; Cantoni, M.; et al. Evidence for spin to charge conversion in GeTe(111). *APL Mater.* **2016**, *4*, 032501. [CrossRef]
41. Zheng, Z.; Su, X.; Deng, R.; Stoumpos, C.; Xie, H.; Liu, W.; Yan, Y.; Hao, S.; Uher, C.; Wolverton, C.; et al. Rhombohedral to Cubic Conversion of GeTe via MnTe Alloying Leads to Ultralow Thermal Conductivity, Electronic Band Convergence, and High Thermoelectric Performance. *J. Am. Chem. Soc.* **2018**, *140*, 2673–2686. [CrossRef] [PubMed]
42. Banik, A.; Vishal, B.; Perumal, S.; Datta, R.; Biswas, K. The origin of low thermal conductivity in $Sn_{1-x}Sb_xTe$: Phonon scattering via layered intergrowth nanostructures. *Energy Environ. Sci.* **2016**, *9*, 2011–2019. [CrossRef]

nanomaterials

MDPI

Article

One-Step Synthesis of Polymeric Carbon Nitride Films for Photoelectrochemical Applications

Alberto Gasparotto [1,2,*], Davide Barreca [2], Chiara Maccato [1,2], Ermanno Pierobon [1] and Gian Andrea Rizzi [1,2,*]

1 Department of Chemical Sciences, Padova University and INSTM, 35131 Padova, Italy; chiara.maccato@unipd.it (C.M.); ermanno.pierobon@unipd.it (E.P.)

2 CNR-ICMATE and INSTM, Department of Chemical Sciences, Padova University, 35131 Padova, Italy; davide.barreca@unipd.it

* Correspondence: alberto.gasparotto@unipd.it (A.G.); gianandrea.rizzi@unipd.it (G.A.R.); Tel.: +39-0498275192 (A.G.); +39-0498275722 (G.A.R.)

Abstract

Over the last decade, polymeric carbon nitrides (PCNs) have received exponentially growing attention as metal-free photocatalytic platforms for green energy generation and environmental remediation. Although PCNs can be easily synthesized from abundant precursors in a powdered form, progress in the field of photoelectrochemical applications requires effective methods for the fabrication of PCN films endowed with suitable mechanical stability and modular chemico-physical properties. In this context, as a proof-of-concept, we report herein on a simple and versatile chemical vapor infiltration (CVI) strategy for one-step PCN growth on porous Ni foam substrates, starting from melamine as a precursor compound. Interestingly, tailoring the reaction temperature enabled to control the condensation degree of PCN films from melem/melon hybrids to melon-like materials, whereas the use of different precursor amounts directly affected the mass and morphology of the obtained deposits. Altogether, such features had a remarkable influence on PCN electrochemical performances towards the oxygen evolution reaction (OER), yielding, for the best performing systems, Tafel slopes as low as ≈65 mV/dec and photocurrent density values of ≈1 mA/cm^2 at 1.6 V vs. the reversible hydrogen electrode (RHE).

Keywords: polymeric carbon nitride films; in situ synthesis; chemical vapor infiltration; Ni foam; oxygen evolution reaction; water splitting

1. Introduction

Polymeric carbon nitrides (PCNs), featuring a pseudo-graphitic structure and comprising C, N, and H (due to residual amino-groups) are a class of attractive and extremely versatile multi-functional materials for a variety of end-uses [1–3]. Among them, melon or melon-like systems, formed by variously condensed 1D chains of amine-linked heptazine units forming H-bonded layers, are the subject of numerous on-going research activities [2,4–6]. These poly-heptazines, often identified with the misnomer 'graphitic carbon nitride' [5–7], are formed by nontoxic abundant elements, can be synthesized from inexpensive precursors, and exhibit a high chemical/thermal stability and appealing characteristics. More specifically, melon-like PCN has a band-gap of ≈2.7 eV, suitable for Vis light absorption, and possesses band edge positions capable of promoting various reactions, such as O_2 and H_2 evolution, CO_2 reduction, and the degradation of organic pollutants [7–11]. Interestingly, PCN reactivity and optoelectronic properties can be tailored by modulating its condensation degree and type/content of defects [2,11,12]. The control of such features, as well as of particle size and morphology [13,14], also allows to circumvent the intrinsic shortcomings of PCNs, including the limited surface area, low electric conductivity, and rapid recombination of photogenerated charge carriers [3,5,8,9].

PCNs are typically synthesized in a powdered form through the thermal polymerization of C,N-containing precursors, such as melamine, urea, or dicyandiamide [3,5,7,15]. Upon annealing in air or inert atmosphere, such compounds undergo a cascade of condensation/de-amination reactions yielding the initial formation of melam, and the subsequent generation of melem (and its oligomers, see below), that then evolves to melon-like PCN [4,5,16]. The modulation of polymerization degree [4,9,17] and amino groups concentration [1,7,18] provides a versatile toolbox for the molecular-level engineering of material reactivity and functional behavior. In this regard, recent works have reported that the photocatalytic performances of melem, its oligomers, and melem/melon hybrids might be superior to melon itself (the archetype PCN material), due to enhanced reactivity and lower electron/hole recombination [4,8–10,17]. Nonetheless, tailoring of PCN surface chemistry and defect content might also favorably impact on wetting properties, a remarkable issue for various (photo)catalytic/(photo)electrochemical reactions [19–22].

Although the synthesis of PCN powders is quite straightforward, the fabrication of films featuring good adhesion to conductive substrates, as required for photoelectrochemical applications, is more challenging and still in the early stages of development due to the difficulties in achieving adequate chemical, physical, and catalytic properties [22–26]. The most common approach to prepare PCN-based films involves the preliminary synthesis of PCN powders and, subsequently, their immobilization on substrates using methods such as spin-coating, dip/drop-casting, doctor-blade, or electrophoretic deposition [15,17,27,28]. Nevertheless, such approaches generally lead to weakly crystalline, inhomogeneous, and scarcely adherent deposits, featuring poor mechanical stability and modest electron transport capacity [12,18,23,27,29,30]. A direct, in situ polymerization route for the one-step growth of PCN films could overcome the above-mentioned problems, also avoiding the drawbacks related to powder processing methods [22,24,30–32]. Approaches for the direct synthesis of PCN films, much less investigated than the two-step routes employed for immobilizing PCN powders, include thermal vapor condensation, microcontact printing, solvothermal routes, electrochemical methods, and chemical vapor deposition (CVD). Up to date, such approaches have been mainly employed for PCN growth on 2D planar substrates, such as fluorine-doped tin oxide (FTO) or indium tin oxide (ITO)-coated glass [11,12,18,22,23,27,29,31,33–35], whereas reports on non-conventional porous supports are almost absent [14,30].

As a step forward in this general context, the present proof-of-concept investigation focuses on the in situ synthesis of PCN films on Ni foam supports through a chemical vapor infiltration (CVI) route. The developed strategy benefits from the molecular-level flexibility and infiltration power of CVD-related techniques to achieve the efficient dispersion of PCN into the porous substrate [12,14,18,36]. Modulation of preparative conditions enabled to tailor PCN loading and morphology, as well as its composition from melem/melon hybrids to melon-like systems. The resulting electrode architectures benefit from the optimal PCN/substrate mechanical adhesion and intimate interfacial contact [15,37], as well as from the high porosity and electrical conductivity of the Ni foam substrate [13,38,39], yielding promising catalytic performances towards the oxygen evolution reaction (OER), the bottleneck of water splitting [28,40]. To the best of our knowledge, no literature works have been reported to date on the OER performances of PCN-based photoelectrodes with different polymerization degrees.

2. Materials and Methods

2.1. Synthesis of Electrode Materials

In the present work, PCN was grown on Ni foam substrates (lateral size ≈1 × 2 cm^2; thickness = 1.7 mm; Ni-4753, RECEMAT BV, Dodewaard, The Netherlands). For complementary chemico-physical and electrochemical characterization, depositions were also carried out on FTO-coated glass supports (lateral size ≈1 × 2 cm^2; overall and FTO layer thickness = 2.2 mm and ≈600 nm, respectively; ≈7 Ω/sq; 735167-1EA, Sigma-Aldrich, Darmstadt, Germany), revealing the facile transferability of the proposed method to diverse support materials. Prior to each deposition, the substrates were subjected to an optimized cleaning procedure [32,41] and fixed on the top of a V-shaped alumina crucible, whose bottom had been previously filled with 100, 200, or 300 mg of pre-grinded melamine powders (99%, Sigma-Aldrich, Darmstadt, Germany). The crucible was hence positioned on a stainless steel susceptor, covered with a second larger alumina vessel, and introduced into a tubular furnace (Carbolite, HST 12/200, Derbyshire, UK) equipped with a quartz tube (inner diameter ≈ 9.5 cm; length of the heated region ≈ 20 cm). PCN deposition (duration = 2.5 h) was carried out at atmospheric pressure under flowing Ar (rate = 3 L/min) in the absence of additional co-reactants, at 500, 550, or 600 °C (heating rate = 5 °C/min). At the end of each experiment, samples were cooled down to room temperature under flowing Ar. The mass of each deposit was measured using a Mettler Toledo (Greifensee, Switzerland) XS105 DualRange microbalance.

2.2. Characterization of Electrode Materials

X-ray diffraction (XRD) measurements were performed in a glancing incidence configuration (θ_i = 1.0°) using a Bruker (Karlsruhe, Germany) AXS D8 Advance Plus diffractometer, equipped with a CuKα X-ray source (λ = 1.54051 Å). The average crystal size was estimated using the Scherrer equation. Optical absorption spectra were collected on a Cary 5E (Varian, Palo Alto, CA, USA) spectrophotometer (spectral bandwidth = 1 nm), operating in transmittance mode on FTO-supported materials. Band-gap (E_G) values were estimated from the corresponding Tauc plots [$(\alpha h\nu)^{1/2}$ vs. hν], assuming indirect allowed transitions [28,42,43]. Field emission-scanning electron microscopy (FE-SEM) analyses were performed by collecting secondary (SE) and backscattered electron (BSE) signals using a Zeiss (Oberkochen, Germany) SUPRA 40VP instrument, at primary beam acceleration voltages of 10–20 kV. X-ray photoelectron spectroscopy (XPS) analyses were performed using a ThermoFisher (Waltham, MA, USA) ESCALAB 250XI+ apparatus, with a monochromatized AlKα X-ray source (hν = 1486.6 eV). Binding energy (BE) values were corrected for charging by setting the adventitious C1s component at 284.8 eV [44]. After a Shirley-type background subtraction, curve fitting was carried out using the XPSpeak 4.1 software [45], using Gaussian–Lorentzian sum functions. Atomic percentages (at.%) were calculated through peak area integration.

2.3. Functional Tests

OER electrochemical tests were carried out both in the dark and under irradiation, using an Autolab PGSTAT204 potentiostat/galvanostat workstation. The prepared samples, a Pt coil, and a Hg/HgO (MMO) electrode were used as working, counter-, and reference electrodes, respectively. The electrolyte was a 0.1 M KOH (Sigma-Aldrich, Darmstadt, Germany) aqueous solution (pH = 12.9). For measurements under irradiation, the working electrode was exposed to a white light LED source (Philips LUMILEDS LXML-PWN1 0120; intensity ≈100 mW/cm^2). Potential values vs. MMO (E_{MMO}) were converted into the reversible hydrogen electrode (RHE) scale using the following equation: $E_{RHE}(V) = E_{WE}(V) + E_{MMO}(V) + 0.0592 \times pH$, where E_{WE} indicates the bias applied to the working electrode. Linear sweep voltammetry (LSV) curves were recorded with a scan rate of 1 mV/s, after activating the samples with cyclic voltammetry (CV) measurements until constant behavior. To assess the material stability, chronoamperometric (CA) analyses were performed at a fixed potential value of 1.5 V vs. RHE under visible light irradiation. Experimentally measured currents were normalized to the electrode geometric area ($\approx$2 cm^2) [46], obtaining current densities in the dark (j_{dark}), under irradiation (j_{light}), and the corresponding photocurrent density values ($j_{light} - j_{dark}$) [47]. Tafel slopes were determined by analyzing the plots of potential vs. log(current density). Electrochemical impedance spectroscopy (EIS) measurements were collected between 50 kHz and 0.5 Hz, with 5 mV sine perturbation.

During the execution of electrochemical experiments, Ni foam-supported specimens featured a marked capillary effect, evidenced by the rise of the electrolytic solution well above the sample immersion level. Because of this phenomenon, likely arising from both the substrate porous structure and the presence of polar groups in the obtained PCN deposits (see below), careful sealing of the electrode contact with epoxy resin turned out to be necessary to ensure a reproducible and accurate measurement setup [19,20,48].

The "coumarin test", a simple and highly sensitive method to assess the relative activity of different catalysts [49], was used to probe the eventual production of •OH radicals by monitoring the formation of the highly photoluminescent species 7-hydroxy-coumarin [49–51]. To this aim, electrochemical experiments were carried out under illumination and constant stirring in a 0.1 M phosphate buffer solution (PBS, pH = 6.9) containing 1 mM coumarin (Sigma-Aldrich, Darmstadt, Germany), using the above-described setup and a saturated calomel reference electrode. Then, Ni foam-supported samples were tested at a fixed bias of 1.6 V vs. RHE, withdrawing, at regular time intervals, aliquots of the solution. The latter were analyzed by collecting fluorescence spectra in the range 340–700 nm using a FLS 1000 fluorimeter (Edinburgh Instruments, Livingston, UK) and adopting the following settings: excitation wavelength/bandwidth = 330/1.5 nm; emission bandwidth = 2.5 nm; optical path = 1 cm.

3. Results and Discussion

A sketch of the setup employed for the one-step growth of PCN deposits on Ni foam supports, along with the main synthesis parameters, is reported in Figure 1a. In a typical experiment, weighted amounts of finely grinded melamine powders (100, 200, or 300 mg) were uniformly spread on the bottom of a V-shaped alumina crucible. Subsequently, pre-cleaned Ni foam and FTO substrates of a suitable size were fit in the upper part of the crucible, ca. 1 cm above the precursor. A second larger crucible was hence positioned face-down on the former one to set up a semi-closed vessel, an important issue to minimize undesired precursor losses and ensure an adequate melamine delivery to the growth surface during the heating stage [52,53]. In this regard, the proper positioning of the top crucible over the bottom one also turned out to be critical to obtain PCN deposits with a reproducible mass. Finally, the double crucible system was positioned on a stainless steel susceptor, introduced into a tubular oven, and heated at the desired temperature in an Ar atmosphere. Under these conditions, precursor vapors infiltrate into the porous Ni foam, and the process is accompanied by de-amination and condensation reactions that yield different products depending on the adopted operating conditions. In this regard, Figure 1b provides a simplified scheme [4,12], reporting approximate temperatures for the progressive conversion of melamine into melam, melem, and melon-based species.

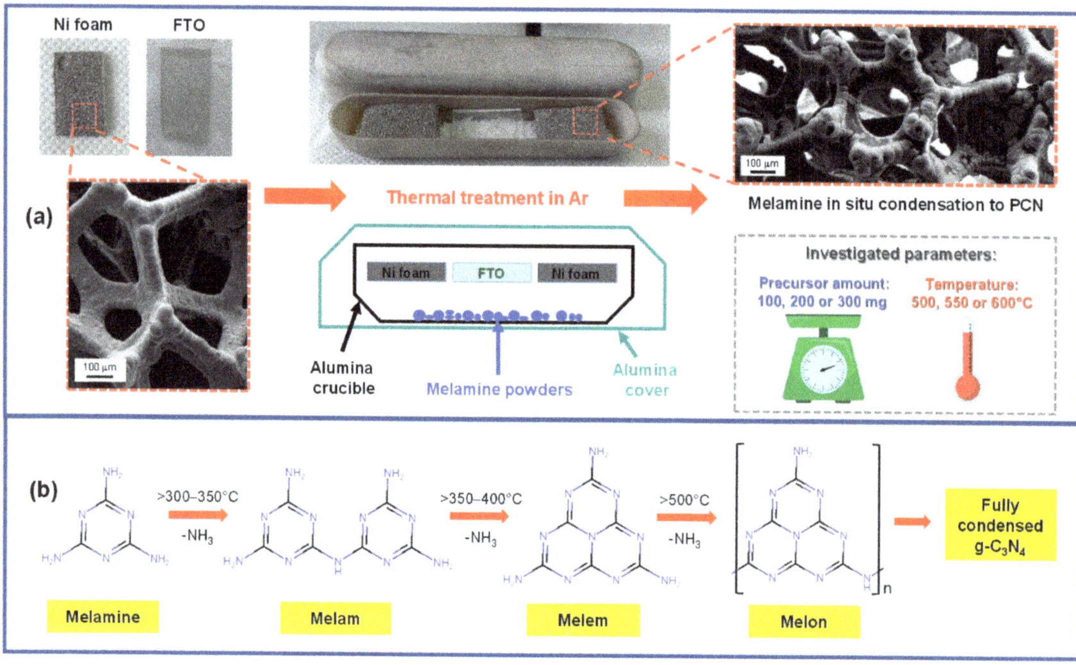

Figure 1. (**a**) Schematic representation of the setup and main process parameters used for PCN deposition on Ni foam and FTO substrates. (**b**) General sketch of melamine main condensation products as a function of temperature. Full condensation should lead to the idealized graphitic carbon nitride (g-C_3N_4) structure [4,54].

Nevertheless, it is worthwhile noticing that different compounds might coexist under certain conditions (e.g., melamine/melem or melem/melon adducts), and various oligomeric species (such as melem dimer, trimer, etc.) can also be formed depending on reaction temperature, time, and atmosphere. In this regard, a more detailed reaction path based on recent literature reports [4,54] is provided in Figure S1.

The composition of precursor residues at the end of the thermal treatment procedure was investigated via Fourier transform infrared (FT-IR) analyses. The corresponding spectra (Figure S2) revealed a broad band between 3000 and 3400 cm^{-1} resulting from the presence of uncondensed amino groups (NH_x, x = 1,2) on PCN ring edges [55,56], beside adsorbed water arising from atmospheric exposure [56,57]. In the 1100–1800 cm^{-1} region, powders resulting from treatments at 550 and 600 °C revealed very similar spectral features, with peaks at 1638, 1568, 1416, 1316, and 1242 cm^{-1} well-matching the stretching modes of C=N/C-N heterocycles of a melon-type PCN material [55,58,59]. The signals at 886 cm^{-1} and 808 cm^{-1} are due to N-H deformation modes and to the breathing of triazine/heptazine units, respectively [13,55,56].

Regarding powders calcined at 500 °C, although the spectrum between 1100 and 1800 cm^{-1} resembled those collected on the 550 and 600 °C samples, the appearance of peaks at 1616, 1466, and 1330 cm^{-1} was consistent with the formation of melem oligomers [4,6,17,59]. Such findings suggest the formation of a melem/melon hybrid material, consistently with a lower melamine condensation degree at 500 °C.

A similar evolution as a function of temperature was also observed for Ni foam- and FTO-supported samples (vide infra). Additionally, tailoring of melamine amount in the reaction vessel (100, 200, or 300 mg) yielded deposits whose mass increased from

(1.0 ± 0.2) mg to (4.0 ± 0.3) mg, a variation accompanied by an appreciable evolution of the corresponding material morphology (see below).

Figure 2a reports the XRD patterns of PCN films grown on FTO at different temperatures. In addition to peaks due to the substrate, the sample obtained at 500 °C exhibited reflections at 2$\vartheta \approx$ 11.0, 12.5, 13.3, 19.7, 22.1, 25.3, 27.1, and 30.4°, which could be assigned to melem oligomers [4,9,60,61]. Nevertheless, the broad bands in the 10–15° and 25–30° regions suggested the co-presence of melon [5,6,53,62], as further supported by XPS results (see below).

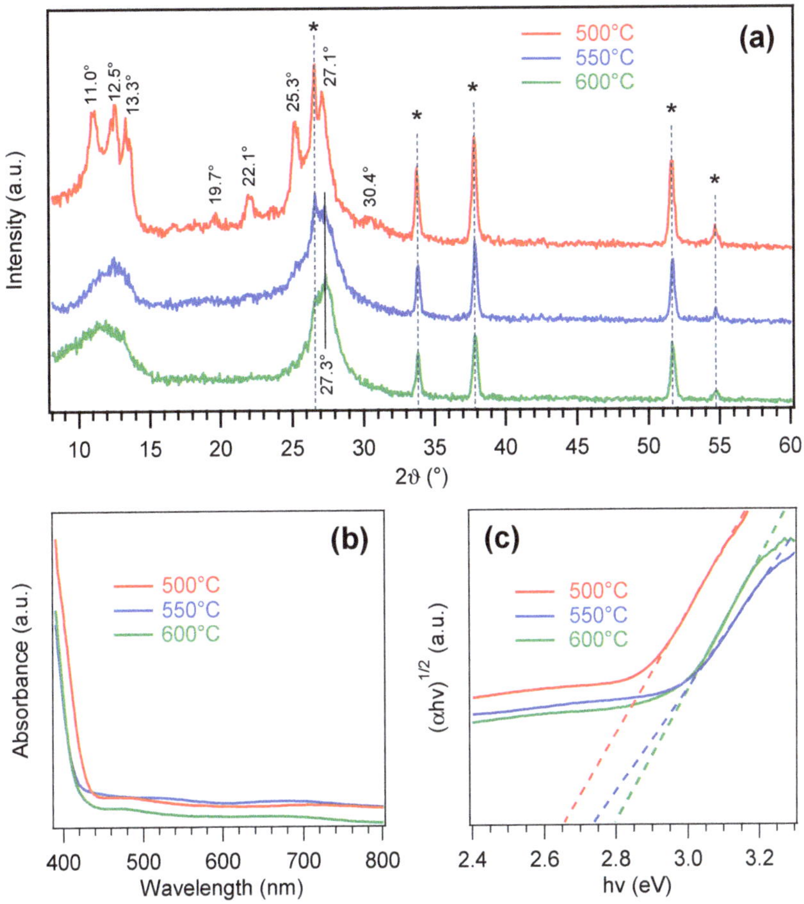

Figure 2. (**a**) XRD patterns of PCN films grown on FTO at different temperatures. Reflections labeled with * are due to the substrate. (**b**) Optical absorption spectra and (**c**) corresponding Tauc plots for the same specimens.

At reaction temperatures of 550 and 600 °C, XRD patterns (relatively similar but markedly different from the 500 °C one) revealed two broad reflections at ≈12.0 and ≈27.0° that, together with the disappearance of several features detected for the 500 °C-grown sample, supported the occurrence of a melon-type material, in tune with a more extensive melamine condensation under harsher temperature conditions [3,5,62]. In particular, peaks at 2$\vartheta \approx$ 12.0 and ≈27.0° can be attributed to melon periodicity along (100) crystallographic planes and (002) interplanar stacking, respectively [5,62,63]. The average crystallite sizes were estimated to be ≈3 nm, for samples grown at 550 and 600 °C, and ≈10 nm, for the specimen synthesized at 500 °C.

The above findings were consistent with the outcomes of optical absorption analysis (Figure 2b). Whereas the 500 °C sample presented an absorption onset at $\lambda \approx 440$ nm, the other two systems featured a blue shift of ≈20 nm. Band-gap determination from Tauc plots (Figure 2c) yielded $E_G \approx 2.65$ eV at 500 °C and ≈2.75–2.80 eV at 550 and 600 °C. Whereas for melon-like materials, E_G values lower than the ones for less-condensed PCN species are typically reported [4,17,53,59], the opposite trend was detected in the present case. Such a finding was traced back to the low nanocrystal size of samples fabricated at the higher temperatures, likely resulting in quantum confinement effects [14,28,30,63].

FE-SEM analyses for Ni foam-supported samples are reported in Figures 3, S3 and S4. The bare Ni foam (Figure 3a) was characterized by a highly porous 3D structure consisting of interconnected branches with a diameter of ≈100 μm. Upon CVI at 500 °C, the Ni foam presented dark regions ascribed to the presence of PCN particles, as revealed by SE micrographs in Figure 3b,c. In this regard, the atomic mass-sensitive BSE image in Figure S3 provides more effective evidence for PCN distribution (revealed by dark-contrast regions) over the Ni foam (bright-contrast). In particular, BSE imaging revealed that even areas where PCN was apparently absent according to the corresponding SE micrograph could be covered by PCN. In this regard, it is worth highlighting that the characterization of Ni foam-supported materials via electron microscopy and spectroscopic techniques might be challenging due to the system multi-scale morphological complexity and tortuous internal structure, limiting collection of emitted electrons to a narrow range of takeoff angles [39,64].

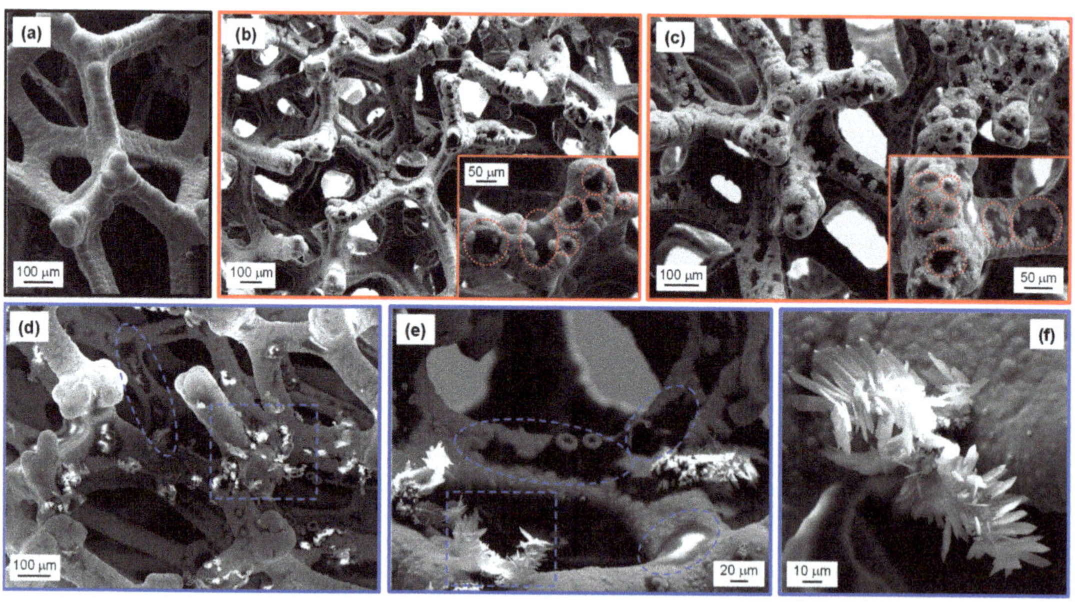

Figure 3. FE-SEM micrographs obtained from secondary electron (SE) signal for: (**a**) bare Ni foam; two representative samples grown at 500 °C from (**b**) 100 mg and (**c**) 300 mg of melamine. In the insets, some island-like PCN particles are highlighted by red circles; (**d**–**f**) different magnification images pertaining to a specimen grown at 550 °C from 300 mg of precursor. In panels (**d**,**e**), selected regions where island- and flake-like PCN structures appear well-evident are enclosed by blue circles and rectangles, respectively.

Basing on the obtained data, specimens synthesized at 500 °C were characterized by PCN particles with an island-like morphology, well-adherent to the underlying Ni foam, an important goal for electrochemical applications [22,23]. A similar morphology has been reported for CVD-grown carbon nitride films on conventional planar substrates [33]. The extent of substrate coverage by PCN islands increased with the precursor amount used during the CVI process (compare Figure 3b,c). Such a result was in line with the different masses of the corresponding deposits (see above).

PCN "islands" were revealed by FE-SEM measurements even for samples fabricated at 550 and 600 °C. Nevertheless, whereas such particles were the only detectable when the synthesis was carried out from 100 mg of precursor, the presence of agglomerated lamellar structures (typical thickness and length ≈ 1–2 and ≈20 μm, respectively) was also noticed when melamine amount was increased, in particular, to 300 mg (see Figure 3d–f). The formation of such "aggregates" (see also Figure S4) at higher reaction temperatures suggested the occurrence of a different growth mechanism under harsher processing conditions. Although the high surface-to-volume ratio of flake-like PCN might appear favorable for the target application, preliminary OER experiments evidenced that such lamellar structures yielded unstable electrochemical performances upon prolonged utilization. Based on the comparison of FE-SEM micrographs collected prior and after functional tests (compare Figure 3d–f with Figure S12a), this effect was traced back to a partial detachment of PCN flakes from the Ni foam substrate. Such a phenomenon, also responsible for the clouding of the electrolytic solution, was particularly critical for specimens synthesized at 600 °C.

The composition of representative samples grown at 500 and 550 °C was investigated by XPS. Survey spectra and surface atomic percentages (see Figure S5 and Table S1) revealed the presence of photoelectron and Auger signals from nitrogen and carbon, along with oxygen and nickel in lower amounts [44,65]. Detection of the latter suggested partial exposure of the Ni foam substrate, in line with FE-SEM results.

Important information was gained from the analysis of C1s and N1s photopeaks in Figure 4, revealing a broadening of both signals on the high BE side for the sample synthesized at 500 °C in comparison to the homologous one obtained at 550 °C. Concerning the latter sample, the C1s peak could be deconvoluted into three components (Figure 4a). Band (i), centered at 284.8 eV, was traced back to adventitious carbon contamination [44,65]. A second relatively weak contribution [(ii); BE = 286.3 eV] was attributed to carbon atoms bonded to uncondensed amino groups (C-NH_x, with x = 1, 2) on heptazine ring edges. The third major band (iii) at 288.1 eV was ascribed to N-C=N carbon atoms in the PCN network [6,14,28,53,66]. For the N1s signal of the 550 °C-grown sample, four contributing components were identified (Figure 4b). The main one [(iv); BE = 398.8 eV] was traced back to bi-coordinated N atoms of PCN moieties (C=N-C, N_{2c}), whereas band (v) at 399.9 eV was attributed to tertiary N centers [N-$(C)_3$, N_{3c}] [6,28,40,53,66,67]. Peak (vi) at 401.0 eV was consistent with -NH_x presence, and peak (vii), at 404.9 eV, was attributed to the excitation of π electrons [28,40,66,68,69].

The C1s and N1s peaks for the sample synthesized at 500 °C are reported in Figure 4c and Figure 4d, respectively. For the former signal, the assignment of components (i), (ii), and (iii) is the same reported above. The additional band (iii*) at 289.3 eV was ascribed to oxidized carbon species involving C-O/C=O moieties on PCN surface [8,70,71]. The N1s peak (Figure 4d), deconvoluted as the one plotted in Figure 4b, revealed a larger relative contribution of band (vi) to the overall signal, in tune with a higher content of uncondensed amino groups at 500 °C. Accordingly, the N/C ratio, calculated excluding the adventitious carbon contribution and based on peak fitting results in Table S2, was ≈1.5 for the specimen grown at 550 °C, in good agreement with the nominal value expected for a melon-like material [59]. Nonetheless, as far as the sample synthesized at 500 °C is concerned, an N/C ratio of ≈1.6 was obtained, supporting the formation of a less condensed melem/melon hybrid system [59]. Although subject of ongoing investigations, several theoretical and experimental studies attribute to edge amine nitrogen atoms a beneficial role in promoting several catalytic reactions, behaving as active centers and improving charge carriers separation, with amino groups acting as hole-stabilizers [60,72–75]. Additionally, together with –OH groups, $-NH_x$ moieties also produce a more hydrophilic surface that facilitates electrolyte penetration into the foam porous structure, accelerates diffusion of hydroxyl-based reactants, and favors the removal of O_2 bubbles, beneficially impacting on OER activity [19,20,22,72].

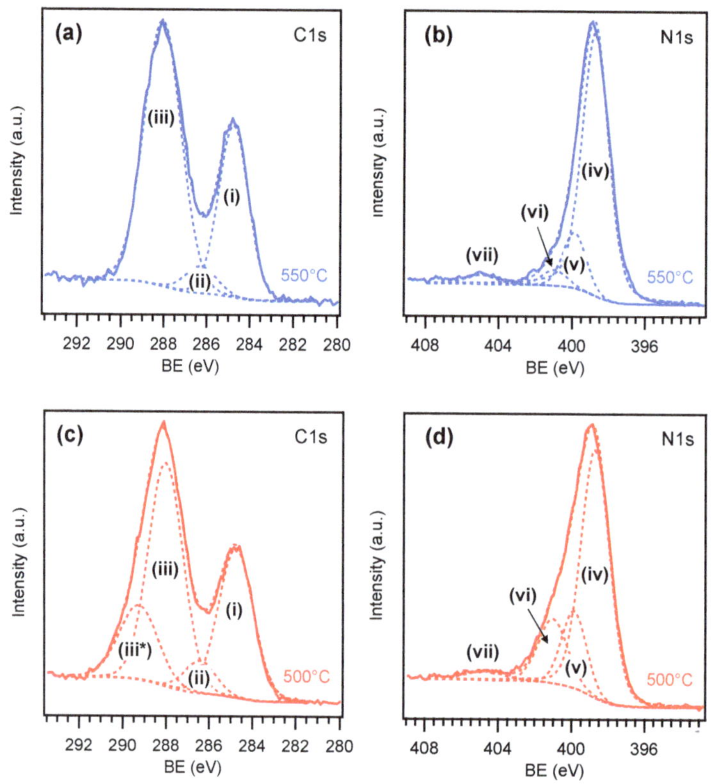

Figure 4. C1s and N1s XPS signals pertaining to two PCN samples grown on Ni foam at (**a**,**b**) 550 and (**c**,**d**) 500 °C.

The electrochemical performances of Ni foam-supported PCN samples were hence tested in 0.1 M aqueous KOH. Results pertaining to the characterization of samples grown at 500, 550, or 600 °C from different melamine amounts are provided in the Supporting Material (see Figures S7–S9 and pertaining discussion), while selected data for the best-performing specimens synthesized at 500 and 550 °C are reported in Figure 5. As a general rule, the samples featured an appreciable current density increase at potentials higher than ≈1.50 V vs. RHE (see LSV curves in Figure 5a and S7a–c), underscoring their OER activity. For each specimen, current densities under irradiation were found to be higher than the corresponding values in the dark, confirming the photoactive nature of PCN deposits [13,15,35]. Analysis of onset potentials (Figure S8) revealed more favorable values for specimens synthesized at 550 and 600 °C, along with a moderate improvement under irradiation. Nonetheless, whereas samples grown at 500 °C were characterized by a retarded onset potential, as well as lower j_{dark} and j_{light} values compared to the systems synthesized at 550 and 600 °C (Figure S9a), they yielded the highest photocurrent density at sufficiently high bias values (see Figure 5b and S9b), provided that an optimal PCN amount was loaded into the Ni foam. Overall, the above functional results are well placed among those reported for various PCN films fabricated either via powder immobilization or direct in situ growth [15,18,27,31,35,76], also taking into account that neither co-catalysts nor sacrificial agents were used in the present case (see also Table S3). The higher photocurrent produced around ≈1.50 V by the 550 °C-grown sample (Figure 5b) can be traced back to an enhanced thermal condensation degree [15,23,76]. Nonetheless, above ≈1.55 V vs. RHE, the sample deposited at 500 °C likely benefits from its superior light-harvesting properties (see above) [10,27].

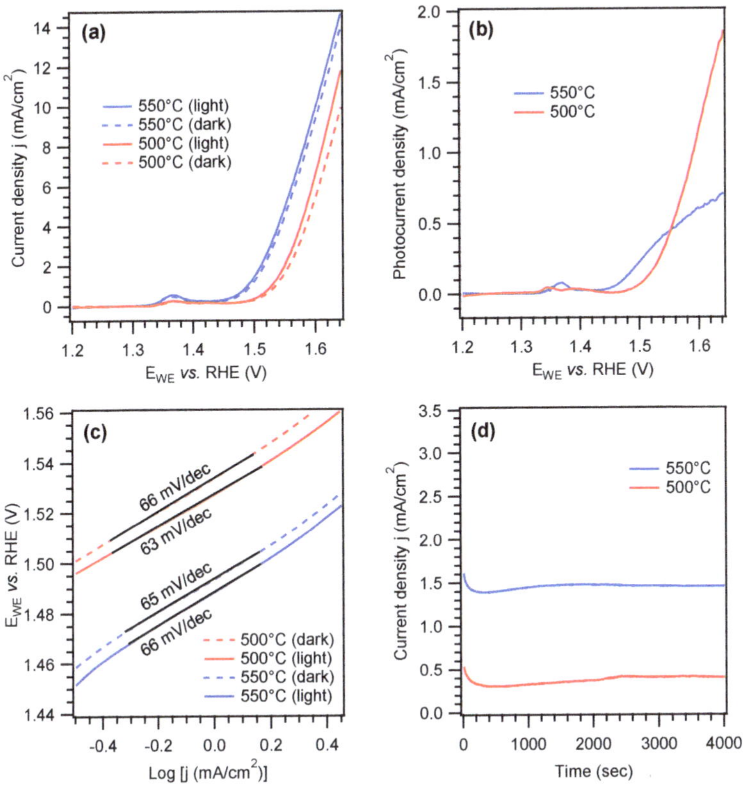

Figure 5. (**a**) LSV anodic scans in 0.1 M KOH collected in the dark (j_{dark}, dotted lines) and under irradiation (j_{light}, continuous lines) on two Ni foam-supported PCN samples grown at 500 °C and 550 °C from 100 and 200 mg of melamine, respectively. The weak anodic peak at ≈1.35 V vs. RHE is due to the NiO ⟶ NiO(OH) reaction involving the uncovered Ni foam surface [13,32,64]. (**b**) Photocurrent density values and (**c**) Tafel plots for the same specimens. In the latter case, continuous black lines mark the fitting of experimental curves. (**d**) Chronoamperometric curves collected under irradiation at 1.5 V vs. RHE.

Tafel slope values for the best-performing samples (≈65 mV/dec, Figure 5c), indicating promising reaction kinetics at the electrode surface, compared favorably with other PCN-based systems reported in the literature [24,40,70]. Under optimized deposition conditions, the proposed CVI approach yielded Ni foam-supported PCN samples with good operational stability, as revealed by the chronoamperometric measurements in Figure 5d. In this regard, FE-SEM analysis after OER experiments revealed that the island-like PCN structures detected prior to functional tests were still clearly evident, without any appreciable alteration (compare Figure 3c and Figure S12b), thus supporting the good morphological and electrochemical stability of PCN deposits and their effective adhesion to the Ni foam.

Overall, for the best-performing systems synthesized at 500 and 550 °C, current density values of 10 mA/cm^2 were achieved under irradiation with an overpotential of ≈395 and ≈370 mV, respectively, whereas the corresponding photocurrents at 1.6 V vs. RHE were ≈1.2 and ≈0.6 mA/cm^2 (Figure 5a,b). As a term of comparison, we have recently reported a two-step method involving: (i) the synthesis of PCN powders by thermal condensation of melamine in Ar atmosphere; (ii) the electrophoretic deposition of the resulting powders on Ni foam [13]. Although experimental conditions for melamine condensation were very similar to the ones adopted herein for the one-step PCN growth, OER performances were appreciably lower, with a photocurrent of ≈0.1 mA/cm^2 at 1.6 V vs. RHE. Additionally, the NiO ⟶ NiO(OH) oxidation peak at 1.35–1.40 V vs. RHE was appreciably more evident, revealing an uneven PCN distribution on the substrate that negatively affected the system conductivity and produced a modest light response [13]. Conversely, the improved functional behavior of the present samples is likely due to the more uniform and effective dispersion of PCN particles into the substrate pores, resulting in an intimate PCN/Ni foam interfacial contact. Nonetheless, considering that PCN has shown an *n*-type behavior (see also Figure S11) and that appreciable current densities are observed only above the potential required for NiO(OH) formation, a co-operative mechanism between the Ni foam and PCN can be hypothesized under illumination, with holes being injected into the electrolyte through NiO(OH) sites, as well as through PCN. In fact, the oxidation potential for Ni^{2+}/Ni^{3+} falls within the PCN energy gap. The ratio between these two paths should depend on the degree of coverage of the Ni foam.

Ni foam-supported samples were also investigated by EIS, both in the dark and under irradiation (see Figure S10a), revealing decreased charge transfer resistance in the latter case, in agreement with LSV curves in Figure 5a. As can be observed, EIS spectra were quite noisy. Such an effect is likely due to the strong O_2 bubbling at high bias values and difficult charge redistribution due to shadowing effects arising from the substrate morphology, also resulting in a relatively modest light response. Mott–Schottky analysis (Figure S10b) revealed that the flat band potential was positioned at ≈1.4 V, in agreement with Figure 5a results.

To supplement (photo)electrochemical data, additional characterization experiments were carried out also on FTO-supported specimens, featuring a much faster light response. To this regard, LSV curves (under dark, light, and chopped conditions), as well as chopped light chronoamperometry and open circuit potential (OCP) scans, were collected (Figure S11). Overall, such data revealed an *n*-type behavior, instead of the amphoteric one often reported for polymeric carbon nitride [28,77]. Additionally, both chronoamperometric and OCP measurements supported the good material electrochemical stability. The slight photocurrent decrease observed in Figure S11b is due to hole accumulation on PCN surface (OCP shifts to higher values after application of a high anodic bias).

As far as the issue of mechanical stability is concerned, on these specimens, allowing to qualitatively assess PCN adhesion by visual inspection, no appreciable detachment/delamination was observed after performing the scotch tape test. As a matter of fact, the deposit could be removed from the substrate only by energy-intensive mechanical scratching.

Ni foam-supported specimens were finally investigated using the "coumarin test", a benchmarked procedure to evaluate a catalyst's ability to produce hydroxyl radicals (•OH) under irradiation. Whereas the formation of such species is highly desired for the degradation of organic aqueous pollutants, we have recently reported that the two-electron mechanism responsible for •OH generation might be competitive with the four-electron process involved in OER [13]. In this regard, results reported in Figure 6 clearly revealed a negligible formation of hydroxyl radicals by the present PCN-based materials, suggesting that such electrocatalysts feature a high selectivity towards water oxidation. In fact, the peak at ≈390 nm is due to coumarin emission, whereas the shoulder at ≈450 nm is due to 7-hydroxy-coumarin (7-OHC), whose formation takes place in the presence of photogenerated •OH [13,49]. Taking into account the much higher photoluminescence quantum yield of 7-OHC, the weak intensity of the corresponding signal reveals a negligible production of hydroxyl radicals.

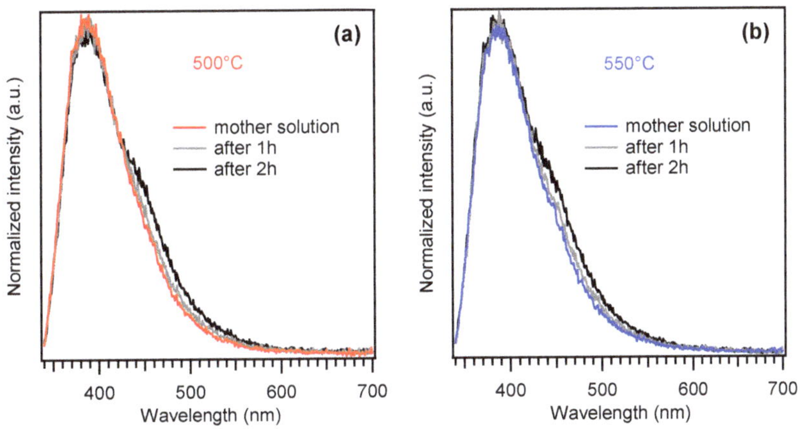

Figure 6. Luminescence spectra obtained from a 1 mM coumarin solution in 0.1 M phosphate buffer after 1 and 2 h of photoelectrochemical work under illumination at a fixed bias of 1.6 V vs. RHE. Panels (**a**,**b**) refer to results obtained on Nifoam-supported PCN samples grown at 500 °C and 550 °C from 100 and 200 mg of melamine, respectively.

4. Conclusions

In this work, we have proposed a cheap and versatile one-step fabrication route to PCN films with tunable compositional, structural, and optical properties. The proposed synthetic approach was optimized on Ni foam substrates yielding *n*-type PCN materials with good adhesion to the substrate, but can also be conveniently extended to other supports. The results of a multi-technique characterization by means of complementary analytical tools indicated that the simple tailoring of preparative conditions, such as reaction temperature and precursor amount, afforded the obtainment of PCN deposits with diversified morphological features and a tunable condensation degree (from melem/melon hybrids to melon-like materials). The modulation of such properties significantly impacted on the OER activity of the corresponding materials yielding, for the best-performing systems, photocurrent density values close to ≈1 mA/cm^2 at 1.6 V vs. RHE and Tafel slopes as low as ≈65 mV/dec. Such results candidate the present materials as potentially promising catalysts for different electrochemical applications. In this regard, additional room for improvement can be provided by performance enhancement strategies, such as co-catalyst deposition, doping, or heterojunction engineering. Efforts in this direction are already under way.

Supplementary Materials: The following are available online at https://www.mdpi.com/article/10.3390/nano15130960/s1, Figure S1: additional details on the reaction mechanism; Figure S2: FT-IR spectra of powders obtained from melamine calcination; Figures S3 and S4: FE-SEM micrographs for representative samples grown at 500 and 550 °C, respectively; Figure S5: XPS survey spectra pertaining two PCN samples grown at 500 and 550 °C; Tables S1 and S2: XPS surface atomic percentages and breakdown of C1s and N1s atomic percentages according to peak fitting deconvolution for the same specimens; Figure S6: Ni2p and O1s XPS signals pertaining two samples grown at 550 and 500 °C; Figure S7: LSV anodic scans collected on Ni foam-supported PCN samples grown at 500 °C, 550 °C, and 600 °C, from different melamine amounts; Figure S8: onset potential under dark and light conditions for PCN samples in Figure S7; Figure S9: current density values in the dark and under irradiation, and photocurrent density values for Figure S7 specimens; Table S3: OER electrochemical performances of selected carbon nitride electrocatalysts reported in the literature; Figure S10: Nyquist plots for Ni foam-supported samples grown at 500 °C and 550 °C and representative Mott-Schottky plot; Figure S11: LSV curves in the dark and under illumination for a FTO-supported sample grown at 550 °C along with chronoamperometric and OCP scans under chopped light for the same material; Figure S12: FE-SEM micrographs recorded after electrochemical tests on Ni foam-supported samples grown at 550 °C and 500 °C. The authors have cited additional references in the Supplementary Materials [78–85].

Author Contributions: Investigation, A.G., E.P., G.A.R. and C.M.; data curation, A.G., C.M., E.P. and G.A.R.; writing—original draft preparation, A.G. and G.A.R.; writing—review and editing, D.B. and C.M.; supervision, A.G., C.M. and G.A.R.; project administration, A.G., G.A.R. and D.B.; funding acquisition, A.G., G.A.R. and D.B. All authors have read and agreed to the published version of the manuscript.

Funding: The present work was financially supported by the National Research Council (Progetti di Ricerca @CNR-avviso 2020-ASSIST), Padova University (PDiSC#02BIRD2023-UNIPD RIGENERA, DOR 2023–2025), INSTM Consortium (TRI.25/013-CIMENTO), and PRIN 2022474YE8 (SCI-TROPHY project; Next Generation EU—Bando PRIN 2022—M4.C2.1.1).

Data Availability Statement: Data supporting this study are available within the article.

Acknowledgments: XPS measurements were carried out in cooperation with Davide Mariotti and Dilli Padmanaban, who are gratefully acknowledged for their technical and scientific support. The authors also thank Mattia Benedet and Florencia Cassola for their support with experimental activities.

Conflicts of Interest: The authors declare no conflicts of interest.

References

1. Tyborski, T.; Merschjann, C.; Orthmann, S.; Yang, F.; Lux-Steiner, M.C.; Schedel-Niedrig, T. Crystal structure of polymeric carbon nitride and the determination of its process-temperature-induced modifications. *J. Phys. Condens. Matter* **2013**, *25*, 395402. [CrossRef] [PubMed]
2. Lotsch, B.V.; Döblinger, M.; Sehnert, J.; Seyfarth, L.; Senker, J.; Oeckler, O.; Schnick, W. Unmasking melon by a complementary approach employing electron diffraction, solid-state NMR spectroscopy, and theoretical calculations—structural characterization of a carbon nitride polymer. *Chem. Eur. J.* **2007**, *13*, 4969–4980. [CrossRef] [PubMed]
3. Alwin, E.; Nowicki, W.; Wojcieszak, R.; Zieliński, M.; Pietrowski, M. Elucidating the structure of the graphitic carbon nitride nanomaterials via X-ray photoelectron spectroscopy and X-ray powder diffraction techniques. *Dalton Trans.* **2020**, *49*, 12805–12813. [CrossRef] [PubMed]
4. Lau, V.W.-H.; Mesch, M.B.; Duppel, V.; Blum, V.; Senker, J.; Lotsch, B.V. Low-molecular-weight carbon nitrides for solar hydrogen evolution. *J. Am. Chem. Soc.* **2015**, *137*, 1064–1072. [CrossRef]
5. Kessler, F.K.; Zheng, Y.; Schwarz, D.; Merschjann, C.; Schnick, W.; Wang, X.; Bojdys, M.J. Functional carbon nitride materials—design strategies for electrochemical devices. *Nat. Rev. Mater.* **2017**, *2*, 17030. [CrossRef]
6. Miller, T.S.; Jorge, A.B.; Suter, T.M.; Sella, A.; Corà, F.; McMillan, P.F. Carbon nitrides: Synthesis and characterization of a new class of functional materials. *Phys. Chem. Chem. Phys.* **2017**, *19*, 15613–15638. [CrossRef]

7. Lau, V.W.-h.; Moudrakovski, I.; Botari, T.; Weinberger, S.; Mesch, M.B.; Duppel, V.; Senker, J.; Blum, V.; Lotsch, B.V. Rational design of carbon nitride photocatalysts by identification of cyanamide defects as catalytically relevant sites. *Nat. Commun.* **2016**, *7*, 12165. [CrossRef]
8. Mo, J.; Wang, N.; Zhang, S.; Chen, X.; Fu, J.; Chen, P.; Liang, Z.; Su, Q.; Li, X. Metal-free g-C_3N_4/melem nanorods hybrids for photocatalytic degradation of methyl orange. *Res. Chem. Intermed.* **2022**, *48*, 3835–3849. [CrossRef]
9. Yang, D.; Li, L.; Xiao, G.; Zhang, S. Steering charge kinetics in metal-free g-C_3N_4/melem hybrid photocatalysts for highly efficient visible-light-driven hydrogen evolution. *Appl. Surf. Sci.* **2020**, *510*, 145345. [CrossRef]
10. Liu, S.; Sun, H.; O'Donnell, K.; Ang, H.M.; Tade, M.O.; Wang, S. Metal-free melem/g-C_3N_4 hybrid photocatalysts for water treatment. *J. Colloid Interface Sci.* **2016**, *464*, 10–17. [CrossRef]
11. Cao, Q.; Kumru, B.; Antonietti, M.; Schmidt, B.V.K.J. Graphitic carbon nitride and polymers: A mutual combination for advanced properties. *Mater. Horiz.* **2020**, *7*, 762–786. [CrossRef]
12. Zou, X.; Sun, Z.; Hu, Y.H. g-C_3N_4-based photoelectrodes for photoelectrochemical water splitting: A review. *J. Mater. Chem. A* **2020**, *8*, 21474–21502. [CrossRef]
13. Benedoue, S.; Benedet, M.; Gasparotto, A.; Gauquelin, N.; Orekhov, A.; Verbeeck, J.; Seraglia, R.; Pagot, G.; Rizzi, G.A.; Balzano, V.; et al. Insights into the photoelectrocatalytic behavior of gCN-based anode materials supported on Ni foams. *Nanomaterials* **2023**, *13*, 1035. [CrossRef] [PubMed]
14. Xing, W.; Tu, W.; Han, Z.; Hu, Y.; Meng, Q.; Chen, G. Template-induced high-crystalline g-C_3N_4 nanosheets for enhanced photocatalytic H_2 evolution. *ACS Energy Lett.* **2018**, *3*, 514–519. [CrossRef]
15. Bian, J.; Li, Q.; Huang, C.; Li, J.; Guo, Y.; Zaw, M.; Zhang, R.-Q. Thermal vapor condensation of uniform graphitic carbon nitride films with remarkable photocurrent density for photoelectrochemical applications. *Nano Energy* **2015**, *15*, 353–361. [CrossRef]
16. Amorin, L.H.; Suzuki, V.Y.; de Paula, N.H.; Duarte, J.L.; da Silva, M.A.T.; Taft, C.A.; de Almeida La Porta, F. Electronic, structural, optical, and photocatalytic properties of graphitic carbon nitride. *New J. Chem.* **2019**, *43*, 13647–13653. [CrossRef]
17. Chu, S.; Wang, C.; Feng, J.; Wang, Y.; Zou, Z. Melem: A metal-free unit for photocatalytic hydrogen evolution. *Int. J. Hydrogen Energy* **2014**, *39*, 13519–13526. [CrossRef]
18. Jia, Q.; Zhang, S.; Gao, Z.; Yang, P.; Gu, Q. In situ growth of triazine–heptazine based carbon nitride film for efficient (photo)electrochemical performance. *Catal. Sci. Technol.* **2019**, *9*, 425–435. [CrossRef]
19. Thi, Q.H.; Man, P.; Huang, L.; Chen, X.; Zhao, J.; Ly, T.H. Superhydrophilic 2D carbon nitrides prepared by direct chemical vapor deposition. *Small Sci.* **2023**, *3*, 2200099. [CrossRef]
20. Ryu, J.; Lee, D.W. Tailoring hydrophilic and hydrophobic microenvironments for gas–liquid–solid triphase electrochemical reactions. *J. Mater. Chem. A* **2024**, *12*, 10012–10043. [CrossRef]
21. Kuila, S.K.; Guchhait, S.K.; Mandal, D.; Kumbhakar, P.; Chandra, A.; Tiwary, C.S.; Kundu, T.K. Dimensionality effects of g-C_3N_4 from wettability to solar light assisted self-cleaning and electrocatalytic oxygen evolution reaction. *Chemosphere* **2023**, *333*, 138951. [CrossRef] [PubMed]
22. Jia, C.; Yang, L.; Zhang, Y.; Zhang, X.; Xiao, K.; Xu, J.; Liu, J. Graphitic carbon nitride films: Emerging paradigm for versatile applications. *ACS Appl. Mater. Interfaces* **2020**, *12*, 53571–53591. [CrossRef] [PubMed]
23. Bian, J.; Xi, L.; Li, J.; Xiong, Z.; Huang, C.; Lange, K.M.; Tang, J.; Shalom, M.; Zhang, R.-Q. C=C π bond modified graphitic carbon nitride films for enhanced photoelectrochemical cell performance. *Chem. Asian J.* **2017**, *12*, 1005–1012. [CrossRef] [PubMed]
24. Fdez-Sanromán, A.; Pazos, M.; Rosales, E.; Sanromán, A. Pushing the operational barriers for g-C_3N_4: A comprehensive review of cutting-edge immobilization strategies. *Catalysts* **2024**, *14*, 175. [CrossRef]
25. Bian, J.; Huang, C.; Zhang, R.-Q. Graphitic carbon nitride film: An emerging star for catalytic and optoelectronic applications. *ChemSusChem* **2016**, *9*, 2723–2735. [CrossRef]
26. Zhu, Y.; He, L.; Ni, Y.; Li, G.; Li, D.; Lin, W.; Wang, Q.; Li, L.; Yang, H. Recent progress on photoelectrochemical water splitting of graphitic carbon nitride (g−CN) electrodes. *Nanomaterials* **2022**, *12*, 2374. [CrossRef]
27. Xie, X.; Fan, X.; Huang, X.; Wang, T.; He, J. In situ growth of graphitic carbon nitride films on transparent conducting substrates via a solvothermal route for photoelectrochemical performance. *RSC Adv.* **2016**, *6*, 9916–9922. [CrossRef]
28. Benedet, M.; Rizzi, G.A.; Gasparotto, A.; Lebedev, O.I.; Girardi, L.; Maccato, C.; Barreca, D. Tailoring oxygen evolution performances of carbon nitride systems fabricated by electrophoresis through Ag and Au plasma functionalization. *Chem. Eng. J.* **2022**, *448*, 137645. [CrossRef]
29. Urakami, N.; Kosaka, M.; Hashimoto, Y. Thermal chemical vapor deposition and luminescence property of graphitic carbon nitride film for carbon-based semiconductor systems. *Jpn. J. Appl. Phys.* **2019**, *58*, 010907. [CrossRef]
30. Dong, F.; Wang, Z.; Li, Y.; Ho, W.-K.; Lee, S.C. Immobilization of polymeric g-C_3N_4 on structured ceramic foam for efficient Visible light photocatalytic air purification with real indoor illumination. *Environ. Sci. Technol.* **2014**, *48*, 10345–10353. [CrossRef]
31. Lu, X.; Liu, Z.; Li, J.; Zhang, J.; Guo, Z. Novel framework g-C_3N_4 film as efficient photoanode for photoelectrochemical water splitting. *Appl. Catal. B* **2017**, *209*, 657–662. [CrossRef]

32. Maccato, C.; Bigiani, L.; Girardi, L.; Gasparotto, A.; Lebedev, O.I.; Modin, E.; Barreca, D.; Rizzi, G.A. Plasma-assisted synthesis of Co_3O_4-based electrocatalysts on Ni foam substrates for the oxygen evolution reaction. *Adv. Mater. Interfaces* **2021**, *8*, 2100763. [CrossRef]
33. Chamorro-Posada, P.; Dante, R.C.; Vázquez-Cabo, J.; Dante, D.G.; Martín-Ramos, P.; Rubiños-López, Ó.; Sánchez-Arévalo, F.M. Experimental and theoretical investigations on a CVD grown thin film of polymeric carbon nitride and its structure. *Diam. Relat. Mater.* **2021**, *111*, 108169. [CrossRef]
34. Chubenko, E.B.; Kovalchuk, N.G.; Komissarov, I.V.; Borisenko, V.E. Chemical vapor deposition of 2D crystallized g-C_3N_4 layered films. *J. Phys. Chem. C* **2022**, *126*, 4710–4714. [CrossRef]
35. Ye, L.; Chen, S. Fabrication and high visible-light-driven photocurrent response of g-C_3N_4 film: The role of thiourea. *Appl. Surf. Sci.* **2016**, *389*, 1076–1083. [CrossRef]
36. Delhaes, P. Chemical vapor deposition and infiltration processes of carbon materials. *Carbon* **2002**, *40*, 641–657. [CrossRef]
37. Shalom, M.; Gimenez, S.; Schipper, F.; Herraiz-Cardona, I.; Bisquert, J.; Antonietti, M. Controlled carbon nitride growth on surfaces for hydrogen evolution electrodes. *Angew. Chem., Int. Ed.* **2014**, *53*, 3654–3658. [CrossRef]
38. Rahman, M.A.; Rahman, M.M.; Song, G. A review on binder-free NiO-Ni foam as anode of high performance lithium-ion batteries. *Energy Storage* **2022**, *4*, e278. [CrossRef]
39. Grdeń, M.; Alsabet, M.; Jerkiewicz, G. Surface science and electrochemical analysis of nickel foams. *ACS Appl. Mater. Interfaces* **2012**, *4*, 3012–3021. [CrossRef]
40. Benedet, M.; Rizzi, G.A.; Gasparotto, A.; Gauquelin, N.; Orekhov, A.; Verbeeck, J.; Maccato, C.; Barreca, D. Functionalization of carbon nitride systems by cobalt and cobalt-iron oxides boosts solar water oxidation performances. *Appl. Surf. Sci.* **2023**, *618*, 156652. [CrossRef]
41. Benedet, M.; Rizzi, G.A.; Lebedev, O.I.; Roddatis, V.; Sada, C.; Wree, J.-L.; Devi, A.; Maccato, C.; Gasparotto, A.; Barreca, D. Advances in photo-assisted seawater splitting promoted by green iron oxide-carbon nitride photoelectrocatalysts. *J. Mater. Chem. A* **2023**, *11*, 21595–21609. [CrossRef]
42. Zhang, D.; Peng, L.; Liu, K.; Garcia, H.; Sun, C.; Dong, L. Cobalt nanoparticle with tunable size supported on nitrogen-deficient graphitic carbon nitride for efficient visible light driven H_2 evolution reaction. *Chem. Eng. J.* **2020**, *381*, 122576. [CrossRef]
43. Ismael, M.; Wark, M. Photocatalytic activity of $CoFe_2O_4$/g-C_3N_4 nanocomposite toward degradation of different organic pollutants and their inactivity toward hydrogen production: The role of the conduction band position. *FlatChem* **2022**, *32*, 100337. [CrossRef]
44. Briggs, D.; Seah, M.P. *Practical Surface Analysis: Auger and X-Ray Photoelectron Spectroscopy*, 2nd ed.; Wiley: New York, NY, USA, 1990.
45. Available online: https://xpspeak.software.informer.com/4.1/ (accessed on 20 February 2020).
46. Benedet, M.; Gallo, A.; Maccato, C.; Rizzi, G.A.; Barreca, D.; Lebedev, O.I.; Modin, E.; McGlynn, R.; Mariotti, D.; Gasparotto, A. Controllable anchoring of graphitic carbon nitride on MnO_2 nanoarchitectures for oxygen evolution electrocatalysis. *ACS Appl. Mater. Interfaces* **2023**, *15*, 47368–47380. [CrossRef]
47. Chen, Z.; Dinh, H.N.; Miller, E. *Photoelectrochemical Water Splitting. Standards, Experimental Methods, and Protocols*; Springer: New York, NY, USA, 2013.
48. Zheng, W.; Liu, M.; Lee, L.Y.S. Best practices in using foam-type electrodes for electrocatalytic performance benchmark. *ACS Energy Lett.* **2020**, *5*, 3260–3264. [CrossRef]
49. Žerjav, G.; Albreht, A.; Vovk, I.; Pintar, A. Revisiting terephthalic acid and coumarin as probes for photoluminescent determination of hydroxyl radical formation rate in heterogeneous photocatalysis. *Appl. Catal. A* **2020**, *598*, 117566. [CrossRef]
50. Nakabayashi, Y.; Nosaka, Y. OH radical formation at distinct faces of rutile TiO_2 crystal in the procedure of photoelectrochemical water oxidation. *J. Phys. Chem. C* **2013**, *117*, 23832–23839. [CrossRef]
51. Nakabayashi, Y.; Nosaka, Y. The pH dependence of OH radical formation in photo-electrochemical water oxidation with rutile TiO_2 single crystals. *Phys. Chem. Chem. Phys.* **2015**, *17*, 30570–30576. [CrossRef]
52. Yuan, Y.; Zhang, L.; Xing, J.; Utama, M.I.B.; Lu, X.; Du, K.; Li, Y.; Hu, X.; Wang, S.; Genç, A.; et al. High-yield synthesis and optical properties of g-C_3N_4. *Nanoscale* **2015**, *7*, 12343–12350. [CrossRef]
53. Thomas, A.; Fischer, A.; Goettmann, F.; Antonietti, M.; Müller, J.-O.; Schlögl, R.; Carlsson, J.M. Graphitic carbon nitride materials: Variation of structure and morphology and their use as metal-free catalysts. *J. Mater. Chem.* **2008**, *18*, 4893–4908. [CrossRef]
54. Lau, V.W.-h.; Lotsch, B.V. A tour-guide through carbon nitride-land: Structure- and dimensionality-dependent properties for photo(electro)chemical energy conversion and storage. *Adv. Energy Mater.* **2022**, *12*, 2101078. [CrossRef]
55. Mo, Z.; She, X.; Li, Y.; Liu, L.; Huang, L.; Chen, Z.; Zhang, Q.; Xu, H.; Li, H. Synthesis of g-C_3N_4 at different temperatures for superior visible/UV photocatalytic performance and photoelectrochemical sensing of MB solution. *RSC Adv.* **2015**, *5*, 101552–101562. [CrossRef]

56. Görmez, Ö.; Yakar, E.; Gözmen, B.; Kayan, B.; Khataee, A. $CoFe_2O_4$ nanoparticles decorated onto graphene oxide and graphitic carbon nitride layers as a separable catalyst for ultrasound-assisted photocatalytic degradation of Bisphenol-A. *Chemosphere* **2022**, *288*, 132663. [CrossRef] [PubMed]
57. Zhou, L.; Lei, J.; Wang, F.; Wang, L.; Hoffmann, M.R.; Liu, Y.; In, S.-I.; Zhang, J. Carbon nitride nanotubes with in situ grafted hydroxyl groups for highly efficient spontaneous H_2O_2 production. *Appl. Catal. B* **2021**, *288*, 119993. [CrossRef]
58. Komatsu, T. The first synthesis and characterization of cyameluric high polymers. *Macromol. Chem. Phys.* **2001**, *202*, 19–25. [CrossRef]
59. Praus, P.; Svoboda, L.; Ritz, M.; Troppová, I.; Šihor, M.; Kočí, K. Graphitic carbon nitride: Synthesis, characterization and photocatalytic decomposition of nitrous oxide. *Mater. Chem. Phys.* **2017**, *193*, 438–446. [CrossRef]
60. Song, X.; Wu, Y.; Pan, D.; Wei, R.; Gao, L.; Zhang, J.; Xiao, G. Melem based multifunctional catalyst for chemical fixation of carbon dioxide into cyclic carbonate. *J. CO2 Util.* **2018**, *24*, 287–297. [CrossRef]
61. Zhang, Q.; Chen, C.; Liu, F.; Zhang, Z.; Fang, X. From pure water to hydrogen peroxide on a novel 2,5,8-triamino-tri-s-triazine (melem)-derived photocatalyst with a high apparent quantum efficiency. *J. Colloid Interface Sci.* **2022**, *625*, 680–691. [CrossRef]
62. Fina, F.; Callear, S.K.; Carins, G.M.; Irvine, J.T.S. Structural investigation of graphitic carbon nitride via XRD and neutron diffraction. *Chem. Mater.* **2015**, *27*, 2612–2618. [CrossRef]
63. Dong, F.; Li, Y.; Wang, Z.; Ho, W.-K. Enhanced visible light photocatalytic activity and oxidation ability of porous graphene-like g-C_3N_4 nanosheets via thermal exfoliation. *Appl. Surf. Sci.* **2015**, *358*, 393–403. [CrossRef]
64. van Drunen, J.; Kinkead, B.; Wang, M.C.P.; Sourty, E.; Gates, B.D.; Jerkiewicz, G. Comprehensive structural, surface-chemical and electrochemical characterization of Nickel-based metallic foams. *ACS Appl. Mater. Interfaces* **2013**, *5*, 6712–6722. [CrossRef] [PubMed]
65. Moulder, J.F.; Stickle, W.F.; Sobol, P.E.; Bomben, K.D. *Handbook of X-Ray Photoelectron Spectroscopy*; Perkin Elmer Corporation: Eden Prairie, MN, USA, 1992.
66. Benedet, M.; Rizzi, G.A.; Barreca, D.; Gasparotto, A.; Maccato, C. XPS analysis of graphitic carbon nitride functionalized with CoO and $CoFe_2O_4$. *Surf. Sci. Spectra* **2023**, *30*, 014004. [CrossRef]
67. Liang, Q.; Li, Z.; Huang, Z.-H.; Kang, F.; Yang, Q.-H. Holey graphitic carbon nitride nanosheets with carbon vacancies for highly improved photocatalytic hydrogen production. *Adv. Funct. Mater.* **2015**, *25*, 6885–6892. [CrossRef]
68. Da Silva, E.S.; Moura, N.M.M.; Coutinho, A.; Dražić, G.; Teixeira, B.M.S.; Sobolev, N.A.; Silva, C.G.; Neves, M.G.P.M.S.; Prieto, M.; Faria, J.L. β-cyclodextrin as a precursor to holey C-doped g-C_3N_4 nanosheets for photocatalytic hydrogen generation. *ChemSusChem* **2018**, *11*, 2681–2694. [CrossRef]
69. Tang, J.; Wang, J.; Tang, L.; Feng, C.; Zhu, X.; Yi, Y.; Feng, H.; Yu, J.; Ren, X. Preparation of floating porous g-C_3N_4 photocatalyst via a facile one-pot method for efficient photocatalytic elimination of tetracycline under visible light irradiation. *Chem. Eng. J.* **2022**, *430*, 132669. [CrossRef]
70. Scattolin, E.; Benedet, M.; Rizzi, G.A.; Gasparotto, A.; Lebedev, O.I.; Barreca, D.; Maccato, C. Graphitic carbon nitride structures on carbon cloth containing ultra- and nano-dispersed NiO for photoactivated oxygen evolution. *ChemSusChem* **2024**, *17*, e202400948. [CrossRef]
71. Bertóti, I.; Mohai, M.; László, K. Surface modification of graphene and graphite by nitrogen plasma: Determination of chemical state alterations and assignments by quantitative X-ray photoelectron spectroscopy. *Carbon* **2015**, *84*, 185–196. [CrossRef]
72. Yang, H.; Zhang, A.; Ding, J.; Hu, R.; Gong, Y.; Li, X.; Chen, L.; Chen, P.; Tian, X. Amino modulation on the surface of graphitic carbon nitride for enhanced photocatalytic hydrogen production. *Carbon* **2024**, *219*, 118841. [CrossRef]
73. Shang, J.; Zhang, K.; Wang, Q.; Jia, S.; Wang, S.; Shen, Z.; Wang, X. Building hydrogen bonds on graphitic carbon nitride for dramatically enhanced ammonia synthesis. *Chem. Eng. J.* **2023**, *452*, 139606. [CrossRef]
74. Cao, S.; Liu, H.; Jia, Z.; Guo, M.; Gao, W.; Ding, Z.; Yang, W.; Chen, L.; Wang, W. Controllable adsorption groups on amine-functionalized carbon nitride for enhanced photocatalytic CO2 reduction. *Chem. Eng. J.* **2023**, *455*, 140746. [CrossRef]
75. Kang, S.; He, M.; Chen, M.; Wang, J.; Zheng, L.; Chang, X.; Duan, H.; Sun, D.; Dong, M.; Cui, L. Ultrafast plasma immersion strategy for rational modulation of oxygen-containing and amino groups in graphitic carbon nitride. *Carbon* **2020**, *159*, 51–64. [CrossRef]
76. Lv, X.; Cao, M.; Shi, W.; Wang, M.; Shen, Y. A new strategy of preparing uniform graphitic carbon nitride films for photoelectrochemical application. *Carbon* **2017**, *117*, 343–350. [CrossRef]
77. Jing, J.; Chen, Z.; Feng, C.; Sun, M.; Hou, J. Transforming g-C_3N_4 from amphoteric to n-type semiconductor: The important role of p/n type on photoelectrochemical cathodic protection. *J. Alloys Compd.* **2021**, *851*, 156820. [CrossRef]
78. Peck, M.A.; Langell, M.A. Comparison of nanoscaled and bulk NiO structural and environmental characteristics by XRD, XAFS, and XPS. *Chem. Mater.* **2012**, *24*, 4483–4490. [CrossRef]
79. Available online: http://srdata.nist.gov/xps (accessed on 18 June 2025).
80. Ohn, S.; Kim, S.Y.; Mun, S.K.; Oh, J.; Sa, Y.J.; Park, S.; Joo, S.H.; Kwon, S.J.; Park, S. Molecularly dispersed nickel-containing species on the carbon nitride network as electrocatalysts for the oxygen evolution reaction. *Carbon* **2017**, *124*, 180–187. [CrossRef]

81. Guo, B.; Tian, L.; Xie, W.; Batool, A.; Xie, G.; Xiang, Q.; Jan, S.U.; Boddula, R.; Gong, J.R. Vertically aligned porous organic semiconductor nanorod array photoanodes for efficient charge utilization. *Nano Lett.* **2018**, *18*, 5954–5960. [CrossRef]
82. Chen, Z.; Wang, H.; Xu, J.; Liu, J. Surface engineering of carbon nitride electrode by molecular cobalt species and their photoelectrochemical application. *Chem. Asian J.* **2018**, *13*, 1539–1543. [CrossRef]
83. Sima, M.; Vasile, E.; Sima, A.; Preda, N.; Logofatu, C. Graphitic carbon nitride based photoanodes prepared by spray coating method. *Int. J. Hydrogen Energy* **2019**, *44*, 24430–24440. [CrossRef]
84. Yan, J.; Wu, H.; Chen, H.; Pang, L.; Zhang, Y.; Jiang, R.; Li, L.; Liu, S. One-pot hydrothermal fabrication of layered β-$Ni(OH)_2$/g-C_3N_4 nanohybrids for enhanced photocatalytic water splitting. *Appl. Catal., B* **2016**, *194*, 74–83. [CrossRef]
85. Mohamed, N.A.; Safaei, J.; Ismail, A.F.; Mohamad Noh, M.F.; Arzaee, N.A.; Mansor, N.N.; Ibrahim, M.A.; Ludin, N.A.; Sagu, J.S.; Mat Teridi, M.A. Fabrication of exfoliated graphitic carbon nitride, (g-C_3N_4) thin film by methanolic dispersion. *J. Alloys Compd.* **2020**, *818*, 152916. [CrossRef]

nanomaterials

MDPI

Article

Uniaxial Magnetization and Electrocatalytic Performance for Hydrogen Evolution on Electrodeposited Ni Nanowire Array Electrodes with Ultra-High Aspect Ratio

Yumu Sako [1], Ryusei Saeki [2], Masamitsu Hayashida [3] and Takeshi Ohgai [3,*]

1 Graduate School of Engineering, Nagasaki University, 1-14 Bunkyo-machi, Nagasaki 852-8521, Japan
2 Graduate School of Engineering, Kyushu University, 744 Motooka, Nishi-ku, Fukuoka 819-0395, Japan
3 Faculty of Engineering, Nagasaki University, Bunkyo-machi 1-14, Nagasaki 852-8521, Japan
* Correspondence: ohgai@nagasaki-u.ac.jp

Abstract: Ni nanowire array electrodes with an extremely large surface area were made through an electrochemical reduction process utilizing an anodized alumina template with a pore length of 320 μm, pore diameter of 100 nm, and pore aspect ratio of 3200. The electrodeposited Ni nanowire arrays were preferentially oriented in the (111) plane regardless of the deposition potential and exhibited uniaxial magnetic anisotropy with easy magnetization in the axial direction. With respect to the magnetic properties, the squareness and coercivity of the electrodeposited Ni nanowire arrays improved up to 0.8 and 550 Oe, respectively. It was also confirmed that the magnetization reversal was suppressed by increasing the aspect ratio and the hard magnetic performance was improved. The electrocatalytic performance for hydrogen evolution on the electrodeposited Ni nanowire arrays was also investigated and the hydrogen overvoltage was reduced down to ~0.1 V, which was almost 0.2 V lower than that on the electrodeposited Ni films. Additionally, the current density for hydrogen evolution at −1.0 V and −1.5 V vs. Ag/AgCl increased up to approximately -580 A/m^2 and -891 A/m^2, respectively, due to the extremely large surface area of the electrodeposited Ni nanowire arrays.

Keywords: anodization; aluminum; nanochannel; electrodeposition; nickel; nanowire; magnetization; electrocatalyst; hydrogen

1. Introduction

Metal nanowires with a high aspect ratio (length/diameter) exhibit shape anisotropy and have a large specific surface area; therefore, they are expected to be used as magnetic and electrocatalytic materials [1–7]. Ferromagnetic metal nanowires such as Fe, Co, and Ni have large magnetic anisotropy based on their shape; therefore, these metals are considered for new hard magnetic materials that can be used in high-density magnetic recording devices. In comparison to non-ferromagnetic metals such as Cu [8], ferromagnetic metals also exhibit excellent electrocatalytic activity for hydrogen evolution in acidic and alkaline baths [9–13]. For example, Shi et al. reported that the Tafel slope for hydrogen evolution on a copper cathode was 160.6 mV dec^{-1} in an aqueous solution containing 1 M Na_2SO_4 and H_2SO_4 [8]. Meanwhile, Mohammadi et al. reported that the Tafel slope for hydrogen evolution on a NiCoP cathode was 49 mV dec^{-1} in an aqueous solution containing 1 M KOH, which was an almost identical value to that of the Pt cathode [10]. Among the ferromagnetic metals, Ni nanowires have been extensively investigated because they are relatively inexpensive and exhibit corrosion resistance [1,4,14].

Metallic nanowires can be synthesized by using an electrodeposition technique from an aqueous solution using a nanochannel template [15], a solvothermal method [16], and a molecular beam epitaxy method [17]. Among them, it is well known that the electrodeposition technique using a nanochannel template has an advantage in terms of cost performance because it enables us to produce numerous metallic nanowires at room temperature under atmospheric pressure [18–22]. As a nanochannel template material for the electrodeposition of metals, anodized aluminum oxide (AAO) templates [23,24] and ion track etched polycarbonate templates [25,26] can be utilized. Zhang et al. discovered that Co nanowires were able to be electrodeposited by utilizing an AC electrodeposition technique into AAO nanochannels with pore diameters of 25 nm, 50 nm, and 75 nm. They revealed that the magnetic squareness of the electrodeposited Co nanowires decreased from 0.924 to 0.525 with increasing the pore diameter up to 75 nm [27].

Some researchers have investigated how nickel and nickel alloy nanowires could be synthesized into nanochannels. For example, Thiem et al. reported that CoNiP nanowire arrays could be fabricated into a polycarbonate nanochannel template by utilizing a potentiostatic deposition technique [28]. They discovered that the squareness of the electrodeposited CoNiP nanowires with a diameter of 100 nm and a length of 3 μm was approximately 0.25. Additionally, Vazquez et al. discovered that Ni nanowires could be produced into alumina nanochannel templates by utilizing a pulsed current electrolysis technique [29]. They revealed that the squareness of the electrodeposited Ni nanowires that were 85 nm in diameter and 4 μm in length was around 0.3. Escrig et al. also discovered that Ni nanowires could be synthesized into alumina nanochannel templates by utilizing a potentiostatic electrodeposition technique [30]. They found that the squareness of the electrodeposited Ni nanowires with a diameter of 100 nm and length of 12 μm was around 0.6. They also discovered that the squareness was improved with increasing the aspect ratio.

Most researchers have attempted to refine the pore diameter of AAO templates to obtain ferromagnetic nanowires with a small diameter, which can improve the hard magnetic performance [31–36]. If the pore diameter of AAO templates is smaller than several tens of nanometers, it will be quite difficult to realize a uniform pore-filling with electrodeposited metallic crystals due to the limitation of metal ion transportation inside the narrow channels. Hence, we attempted to increase the pore length to realize an extremely large aspect ratio in the present study. However, if the anodic oxidation time is longer than several tens of hours at room temperature, the AAO films will be redissolved in the acidic aqueous solution [37]. Therefore, it is quite difficult to make an AAO template with an extremely large pore length. If we can keep the solution temperature lower than 10 degrees Celsius by utilizing a cool incubator system, the redissolution of the AAO film will be prevented and an AAO template with an extremely large pore length could be synthesized. By utilizing an AAO template with a large aspect ratio, the specific surface area of electrodeposited Ni nanowire arrays will be significantly increased. Hence, the electrocatalytic activity for hydrogen evolution on the electrodeposited Ni nanowire arrays will be also improved.

Therefore, in the present study, we developed AAO templates with a pore length (L) of 320 μm, pore diameter (D) of 100 nm, and pore aspect ratio (L/D) of 3200 by keeping the solution temperature lower than ten degrees Celsius by utilizing a cool incubator system during the anodic oxidation process. Subsequently, Ni nanowire arrays with an ultra-high aspect ratio were electrochemically synthesized using the developed AAO templates. The preferential crystal orientation, uniaxial magnetization performance, and electrocatalytic activity for hydrogen evolution were also investigated.

2. Materials and Methods

A commercially available aluminum rod (The Nilaco Corporation, Tokyo, Japan, item number: AL-012659, purity: 99%) with a diameter of 10 mm was mechanically polished in the cross-section area using sandpaper (#1500). Subsequently, to achieve a mirror-like surface finish, electrochemical polishing was applied to the cross-section of the aluminum rod in an ethanol solvent with 20 vol.% perchloric acid (Wako Pure Chemical Industries, Ltd., Osaka, Japan, 60–62 mass%). During the mirror-like surface finishing process, the cell voltage was maintained at 50 V for 2 min. Anodic oxidation was then performed in an aqueous solution containing 0.6 mol/L oxalic acid. During the anodic oxidation process, the cell voltage was kept at 90 V for 24 h to generate an anodic oxide layer with a nanochannel structure. After the anodic oxidation process, the thin disc layer of aluminum oxide was exfoliated from the cross-section of the metallic aluminum rod in an ethanol solvent with 50 vol.% perchloric acid to produce an anodized aluminum oxide (AAO) nanochannel template. Subsequently, the AAO nanochannel template was chemically etched in an aqueous solution containing 8 wt.% phosphoric acid at room temperature (25 °C) to remove the barrier layer [33]. After the chemical etching process, a conductive metallic copper layer was sputtered on one side of the AAO nanochannel template to apply as a working electrode for growing Ni nanowire arrays. The nanochannel diameter and length were determined by using a field-emission scanning electron microscope (FE-SEM, JSM-7500FA, JEOL, Tokyo, Japan) and a digital micrometer (MDC-25MX, Mitsutoyo, Yokohama, Japan), respectively.

The nanowires were electrochemically synthesized into the AAO nanochannel template in an aqueous solution containing 0.5 M nickel sulfate and 0.4 M boric acid. The solution pH and temperature were kept at 4.0 and 40 °C, respectively. Before the electrochemical growing process, the AAO nanochannel templates were immersed in an aqueous solution containing 0.5 M nickel sulfate and 0.4 M boric acid under reduced pressure to fill the solution into the nanochannels. A metallic nickel plate was used as a soluble anode while a Ag/AgCl electrode was used as a reference electrode. During the potentiostatic electrochemical reduction, the growing process of Ni nanowires was monitored by measuring the time dependence of the cathodic current. After the electrochemical growing process, the AAO nanochannel templates were removed by dissolving in an aqueous solvent with 5 M NaOH to separate the Ni nanowire arrays from the AAO templates. Subsequently, the morphology of the recovered Ni nanowires was investigated by utilizing a field-emission scanning electron microscope (FE-SEM, JSM-7500FA, JEOL, Tokyo, Japan) and a transmission electron microscope (TEM, JEM-2010-UHR, JEOL Ltd., Tokyo, Japan).

The preferential crystal orientation of the Ni nanowire arrays was analyzed by using X-ray diffraction (XRD, MiniFlex 600-DX, Rigaku Corp., Tokyo, Japan) patterns and electron diffraction (ED) patterns. The uniaxial magnetization performance of the Ni nanowires was evaluated using a vibrating sample magnetometer (VSM, TM-VSM1014-CRO, Tamakawa Co. Ltd., Sendai, Japan) by increasing the magnetic field up to 10 kOe at room temperature. Electrocatalytic activity for hydrogen evolution on the Ni nanowire array electrodes was investigated using a linear sweep voltammogram (LSV, Electrochemical Measurement System, HZ-7000, Hokuto Denko Corp., Tokyo, Japan) in an aqueous solvent with 0.025 M sulfuric acid at 40 °C. During the LSV measurement, a gold wire was used as an insoluble anode while a Ag/AgCl electrode was used as a reference electrode. The cathode potential was linearly swept in the range from −0.5 V to −1.5 V vs. Ag/AgCl.

3. Results and Discussion

3.1. Fabrication of Anodized Aluminum Oxide (AAO) Template

Figure 1 shows SEM images of the bottom surface (a), cross-section (b), and top surface (c) of an AAO nanochannel template that was anodically oxidized at 90 V for 24 h. During the anodic oxidation process, the bath temperature was kept at 10 °C to prevent the redissolution of the AAO film in the aqueous solution containing oxalic acid. According to these SEM images, it was confirmed that the AAO nanochannel template had a through-hole structure. Therefore, in the present study, the nanochannel length was assumed to be the same as the AAO film thickness. The film thickness was determined by using a digital micrometer as described in the experimental section. The exfoliated AAO template had a nanochannel diameter (D) of ~100 nm and a nanochannel length (L) of ~320 μm. Hence, the aspect ratio (L/D) of the nanochannels was estimated to be ~3200.

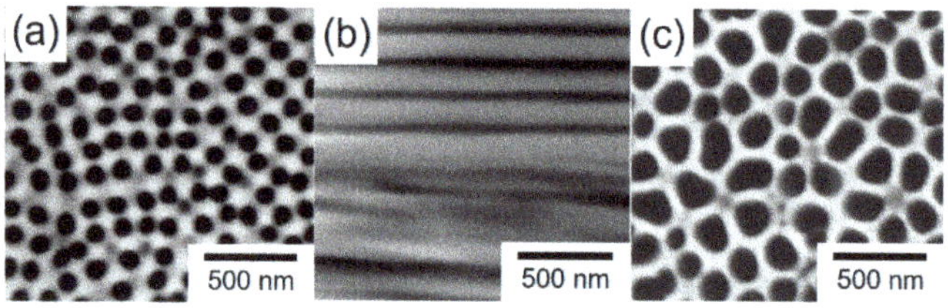

Figure 1. SEM images of an anodic oxidation coating exfoliated from a metallic aluminum surface: (**a**) top view of the film, (**b**) cross-section view of the film, and (**c**) bottom view of the film. The anodization cell voltage was kept at 90 V for 24 h.

3.2. Electrochemical Growth of Ni Nanowires in the AAO Nanochannels

Figure 2 shows the cathodic polarization behavior for Ni electrochemical growing from an acidic aqueous solvent with nickel sulfate and boric acid. During the measurement, a copper foil was used as a cathode while a nickel plate was used as a soluble anode. In addition, a Ag/AgCl electrode was also used as a reference electrode. The potential sweep rate was fixed at 50 mV s^{-1}. The equilibrium potential for Ni/Ni^{2+} can be estimated by the following Nernst Equation (1):

$$E_{Ni}^{eq} = E_{Ni}^{0} + \frac{RT}{2F} ln \frac{\left[Ni^{2+}\right]}{\left[Ni^{0}\right]} \quad (1)$$

where E_{Ni}^{eq}, E_{Ni}^{0}, F, and R represent the equilibrium potential for Ni/Ni^{2+}, the standard electrode potential for Ni/Ni^{2+}, the Faraday constant (96,485 C mol^{-1}), and the gas constant (8.31 J K mol^{-1}), respectively, and T, $[M^{n+}]$, and $[M^0]$ denote the absolute temperature, the activity of metal ions, and the activity of electrodeposited metal, respectively.

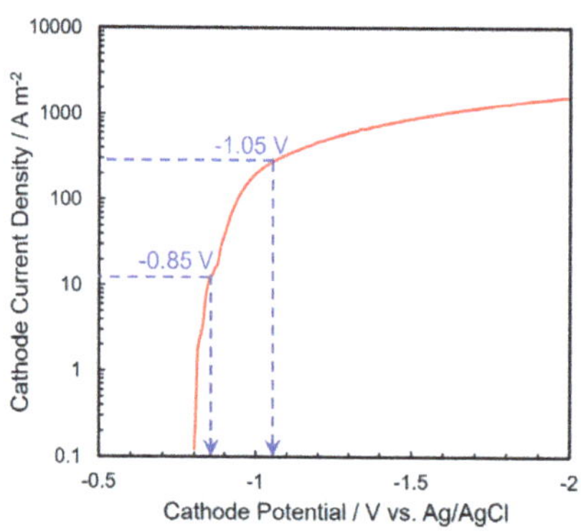

Figure 2. Cathode polarization curve for the electrodeposition of Ni from an acidic aqueous solution.

According to the above Nernst equation, E_{Ni}^{eq} can be determined to be −0.436 V vs. Ag/AgCl at 40 °C bath temperature. As depicted in the cathodic polarization behavior (Figure 2), however, the cathodic current rises up at a potential of around −0.80 V vs. Ag/AgCl. Usually, the electrochemical reduction of iron group metal ions such as Fe^{2+}, Co^{2+}, and Ni^{2+} proceeds accompanying a kinetic overpotential based on the multi-step electrochemical reduction mechanism via hydroxide ions as per the following Equations (2)–(4) [38]:

$$Ni^{2+} + OH^{-} = NiOH^{+} \quad (2)$$

$$NiOH^{+} + e^{-} = [NiOH]_{ad} \quad (3)$$

$$[NiOH]_{ad} + e^- = Ni + OH^- \tag{4}$$

Therefore, this increase in the cathodic current (Figure 2) seems to correspond to the electrodeposition current of Ni^{2+} ions. According to Figure 2, the following Tafel Equation (5) can be applied in the cathode current density range from 1 A/m^2 to 100 A/m^2.

$$\eta_c = a + b \log i_c \tag{5}$$

where η_c, i_c, a, and b represent the cathodic overpotential, cathode current density, Tafel constant, and Tafel slope, respectively. The Tafel slope, b, can be also described by the following Equation (6):

$$b = \frac{2.303RT}{\alpha zF} \tag{6}$$

Here, α and z correspond to the charge transfer coefficient and number of electrons exchanged, respectively. If the electrodeposition process of Ni is dominated by the above Equation (4), α and z can be assumed as 0.5 and 1, respectively. Hence, the Tafel slope, b, can be estimated as 0.124. This theoretical value corresponds well with the experimental value ($b = 0.13$) which was determined from the Tafel plot in Figure 2 in the cathode current density range from 1 A/m^2 to 100 A/m^2.

Moreover, with increasing the cathode current density to more than 200 A/m^2, the cathode potential polarized significantly because the rate-determining stage shifted from the charge-transfer stage (Equation (4)) to the mass-transfer stage of Ni^{2+} ion migration and diffusion. In this cathode current density range, the cathode current efficiency will be decreased due to hydrogen evolution which is caused by the decomposition of water solvent. This side reaction will result in the formation of powder-like or dendrite-like Ni deposits. Furthermore, the Ni^{2+} ion diffusion coefficient in AAO nanochannels will be smaller than that on a flat cathode. Therefore, in the present study, the cathode potential for the electrochemical growth of Ni nanowire arrays was fixed to the range from −0.85 V to −1.05 V vs. Ag/AgCl.

Figure 3a depicts the time dependence of the cathode current during the Ni nanowires growth utilizing a potentiostatic electrodeposition technique. At the first stage, the observed current value was almost constant due to the Ni nanowires' homogeneous growth in the AAO nanochannels. After the first stage, the cathode current value increased rapidly at the electrodeposition time of several thousand seconds. Usually, the cathodic current will be in proportion to the cathode surface area. Hence, this drastic enhancement in the cathode current seemed to be induced by the formation of film-like Ni deposits on the outside of the nanochannels. Furthermore, with an increase in the electrochemical reduction time, the cathode current asymptotically reached a stable value. At this final stage, the AAO template surface seemed to be completely covered with the electrodeposited Ni film and the cathode surface area seemed to reach a maximum value. However, in the final stage at −1.05 V (red line in Figure 3a), the cathode current slightly decreased. This seems to be owing to the formation of cracks, which can be induced from the internal stress of the electrodeposited Ni film. According to the duration of the first stage in Figure 3a, the crystal growth rate of electrodeposited Ni nanowires was determined by dividing the length of the AAO nanochannels by the growth time of the Ni nanowires in the AAO nanochannels. Figure 3b depicts the cathode potential dependence on the growth rate of Ni nanowires. With increasing the cathodic overvoltage, the growth rate of Ni nanowires decreased exponentially. Ni nanowires were electrodeposited at a growth rate ranging from 20 to 70 nm/s, which was significantly faster than that achieved by a conventional sputtering process (approximately 1 nm/s) [39].

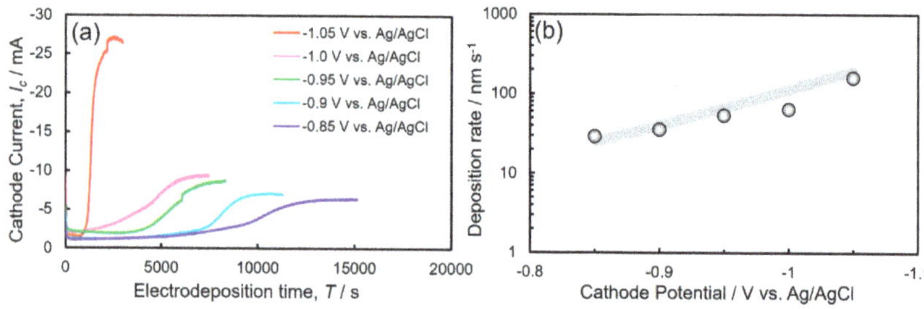

Figure 3. (**a**) Time dependence of cathode current during the electrodeposition of Ni nanowires from the sulfuric acid bath. (**b**) Effect of cathode potential on the deposition rate of Ni nanowires.

3.3. Crystal Texture of Electrodeposited Ni Nanowires

Figure 4a depicts an SEM image of an electrochemically grown Ni nanowire array that was recovered from a dissolved AAO template. According to the SEM image, the numerous Ni nanowires aggregated each other due to their ultra-high aspect ratio of more than 3000.

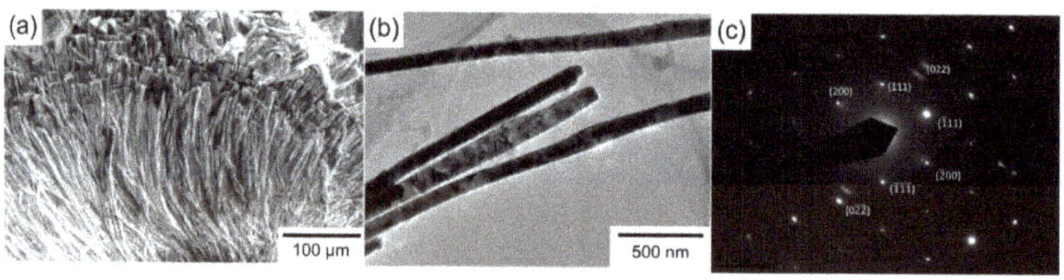

Figure 4. (**a**) SEM image of electrodeposited Ni nanowire array. (**b**) TEM image (brightfield image) of electrodeposited Ni nanowires. (**c**) Electron diffraction patterns of electrodeposited Ni nanowires.

Figure 4b,c show a TEM brightfield image and an ED pattern of the electrochemically grown Ni nanowires. The TEM image reveals that the diameter of the electrodeposited Ni nanowires is around 100 nm, which is almost identical to the pore diameter of the AAO template. It was also confirmed that the Ni nanowires that were grown at the potential of −0.90 V vs. Ag/AgCl consisted of an fcc-Ni phase due to the ED spot pattern arrangement. In particular, spots deriving from fcc-Ni (111) were clearly observed.

Figure 5a depicts the XRD profiles of Ni thin films that were electrochemically grown at cathode potentials of −0.85 V, −0.90 V, −0.95 V, −1.00 V, and −1.05 V. The electrodeposition conditions such as bath composition and film thickness for the Ni thin films were the same as those for the Ni nanowires. In these XRD patterns, the peak (2θ = 44.5°) that was derived from fcc-Ni (111) was observed clearly when the sample was electrodeposited at a noble potential range, while the peak (2θ = 51.8°) that was derived from fcc-Ni (200) was also observed apparently as the sample was grown at a less noble potential range [40]. In contrast, according to the XRD pattern that was obtained from the electrodeposited Ni nanowire arrays, the fcc-Ni (111) was strongly oriented as shown in Figure 5b. This tendency corresponds well to the result which was obtained from the electron diffraction pattern during the TEM observation (Figure 4c).

On the basis of the above XRD patterns, the crystal texture coefficients $TC_{(hkl)}$ of the electrodeposited Ni nanowire arrays were calculated using the following Harris equation [41,42]:

$$TC_{(hkl)} = \frac{\frac{I^i_{hkl}}{I^0_{hkl}}}{\frac{1}{N} \times \sum_{j=1}^{N} \left(\frac{I^j_{hkl}}{I^0_{hkl}} \right)} \quad (7)$$

where I^0 denotes the intensity of each peak in the XRD profile that was obtained from a standard Ni powder, I^i denotes the intensity of each peak in the XRD profile that was obtained from the electrodeposited Ni sample, $(h\,k\,l)$ corresponds to each lattice plane of the Ni crystal, and N is the number of observed diffraction profiles.

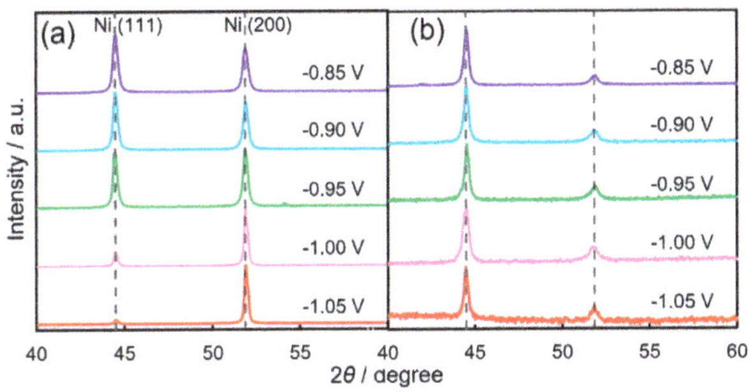

Figure 5. X-ray diffraction patterns of Ni thin films (**a**) and Ni nanowire arrays (**b**) that were electrodeposited at −0.85 V, −0.90 V, −0.95 V, −1.00 V, and −1.05 V.

Figure 6 depicts the cathode potential dependence on the texture coefficients $TC_{(111)}$ of electrodeposited Ni thin films (a) and electrodeposited Ni nanowire arrays (b). The $TC_{(111)}$ of the Ni thin films decreased down to 0.046 with shifting the potential to −1.05 V vs. Ag/AgCl. This is because the overvoltage was increased by shifting the potential to a less noble direction during the electrodeposition. This result is consistent with Pangarov's theory [43–45]. In contrast, the $TC_{(111)}$ of the Ni nanowires kept a stable value higher than 1.4 over the wide cathode potential range.

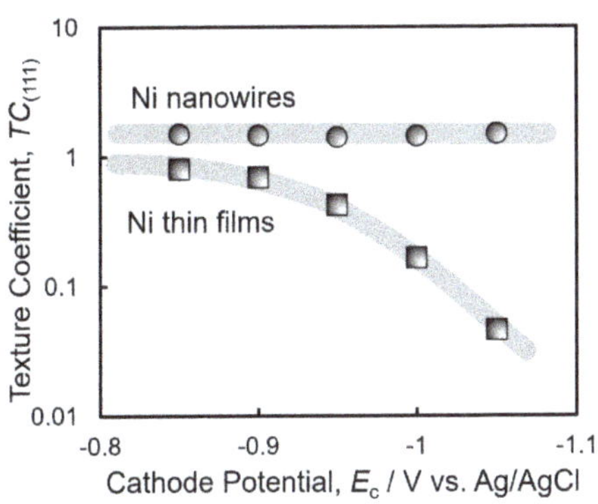

Figure 6. Effect of cathode potential on the texture coefficients $TC_{(111)}$ of Ni thin films and Ni nanowire arrays.

3.4. Magnetic Properties of Electrochemically Grown Ni Nanowires

Figure 7 depicts the magnetic hysteresis loops of Ni thin films and Ni nanowire arrays that were electrodeposited in the nanochannels of AAO films. The magnetic hysteresis loops were obtained when an external magnetic field was applied in an in-plane direction (dashed line) and perpendicular direction (solid line) to the thin film and AAO film plane. The perpendicular direction corresponds to the axial direction of Ni nanowires. As shown in the blue solid lines in Figure 7b–f, Ni nanowires were magnetized spontaneously in the axial direction and exhibited uniaxial magnetic anisotropy. Thus, the shape magnetic anisotropy which is based on the high aspect ratio of the Ni nanowires was confirmed.

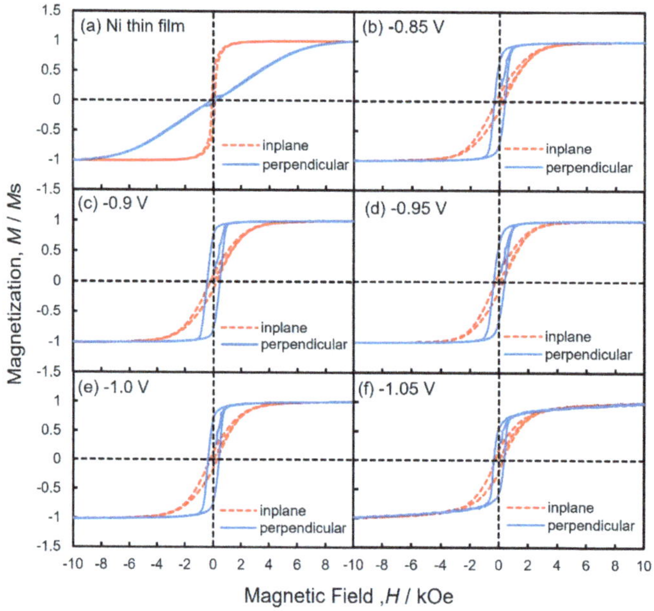

Figure 7. Magnetic hysteresis loops of Ni thin films and Ni nanowire arrays that were electrodeposited at each cathode potential. Magnetic field was applied in the in-plane direction and perpendicular to the film plane.

Figure 8 shows the cathode potential dependence on the coercivity (a), squareness (b), and pore filling ratio (c) of the electrodeposited Ni nanowire arrays. As shown in Figure 8a, the coercivity of the electrodeposited Ni thin films was only 45 Oe [46]. On the contrary, the coercivity of the electrodeposited Ni nanowire arrays increased up to 502 Oe. On the basis of the XRD patterns in Figure 5b, the electrodeposited Ni nanowire arrays exhibited a preferential crystal growth direction in fcc-Ni [111] which corresponds to the easy magnetization direction of fcc-Ni crystal. Therefore, the enhancement in the coercivity of electrochemically grown Ni nanowires with the preferential crystal growth direction in [111] seems to be caused by the consistency between the magneto-crystalline anisotropy and shape magnetic anisotropy of Ni nanowires. As shown in Figure 8b, the squareness of electrochemically grown Ni nanowires increased with decreasing the cathodic overpotential and reached up to 0.84 at the potential of -0.85 V vs. Ag/AgCl. Escrig et al. revealed that the squareness of electrodeposited Ni nanowires increased with increasing the aspect ratio and reached around 0.7 in a sample with an aspect ratio of 240 (50 nm in diameter and 12 μm in length) [30]. Neetzel et al. synthesized electrochemically grown Fe nanowire arrays with a diameter (D) of 85 nm, average length (L) of 60 μm, and average aspect ratio (L/D) of approximately 706 [36]. They revealed that the electrodeposited Fe nanowire arrays exhibited squareness (Mr/Ms) and coercivity (Hc) values of approximately 0.59 and 550 Oe, respectively. Fan et al. reported that they electrodeposited Co/Cu multilayered nanowires with an average diameter (D) of 60 nm, a length (L) of 7 μm, and an aspect ratio (L/D) of approximately 117 [47]. They discovered that the coercivity (Hc) and squareness (Mr/Ms) of the multilayered nanowires were approximately 570 Oe and 0.15, respectively. Usually, electrodeposited iron group metal films are strongly affected by a demagnetizing field when an external magnetic field is applied in the perpendicular direction to the film surface. On the contrary, for cylindrical iron group metal nanowires with a large aspect ratio, the demagnetization factor becomes small when an external magnetic field is applied in the axial direction. The demagnetizing field, H_d, can be given by the following Equation (8):

$$H_d = \left(\frac{N_d}{\mu_0}\right) \times M \tag{8}$$

where N_d represents the demagnetization factor, μ_0 corresponds to the magnetic permeability $\left(4\pi \times 10^{-7}[\text{H/m}]\right)$, and M means the magnetization. Additionally, N_d can be expressed by Equation (9) as a function of the aspect ratio, $k = L/D$ (L: length of nanowire; D: diameter of nanowire):

$$N_d = \frac{1}{k^2-1}\left\{\frac{k}{\sqrt{k^2-1}}\ln\left(k+\sqrt{k^2-1}\right)-1\right\} \tag{9}$$

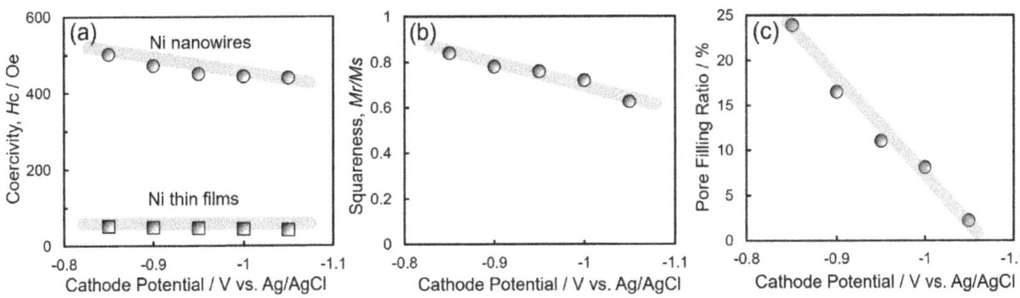

Figure 8. Effect of cathode potential on coercivity (**a**), squareness (**b**), and pore filling ratio (**c**) of electrodeposited Ni nanowire arrays.

According to the above Equation (9), the demagnetization factor N_d will decrease down to 1.7×10^{-5} from 4.3×10^{-4} when the aspect ratio k increases up to 600 from 100. Furthermore, N_d will decrease down to 8.6×10^{-7} when k increases up to 3000. Thus, the demagnetization factor decreases as the aspect ratio increases. On the basis of Equation (8), the demagnetizing field will decrease with a decrease in the demagnetization factor. Therefore, the nanowires will be magnetized easily in the axial direction when the aspect ratio increases. This uniaxial magnetization performance will enhance the hard magnetic properties such as squareness and coercivity. As depicted in Figure 8b, the squareness improved as the potential shifted to the electrochemical noble region. This improvement in the squareness seems to be attributed to the increase in the pore filling ratio as shown in Figure 8c. If the cathodic overvoltage is decreased, the throwing power will be improved because the rate-limiting process during the electrodeposition will become the charge transfer process from the mass transfer process, such as the migration of metal ions. Therefore, the average aspect ratio of electrodeposited Ni nanowires seems to be enhanced by decreasing the cathodic overvoltage.

3.5. Electrocatalytic Performance for Hydrogen Evolution on the Electrodeposited Ni Nanowire Arrays

Figure 9 depicts the polarization curves for hydrogen evolution reaction on the electrodeposited Ni thin films and Ni nanowire arrays in an aqueous solvent with sulfuric acid. According to Figure 9, the hydrogen evolution reaction occurred at the potential of approximately −0.5 V vs. Ag/AgCl on the Ni thin film electrode whereas it occurred at the potential of approximately −0.3 V vs. Ag/AgCl on the Ni nanowire array electrodes. Hence, the minimum hydrogen overvoltage is estimated to be around 0.3 V on the electrodeposited Ni thin films whereas it is estimated to be around 0.1 V on the Ni nanowire array electrodes. This reduction in the minimum hydrogen overvoltage seems to be caused by the enhancement of the specific surface area on the electrodeposited Ni nanowire array electrodes.

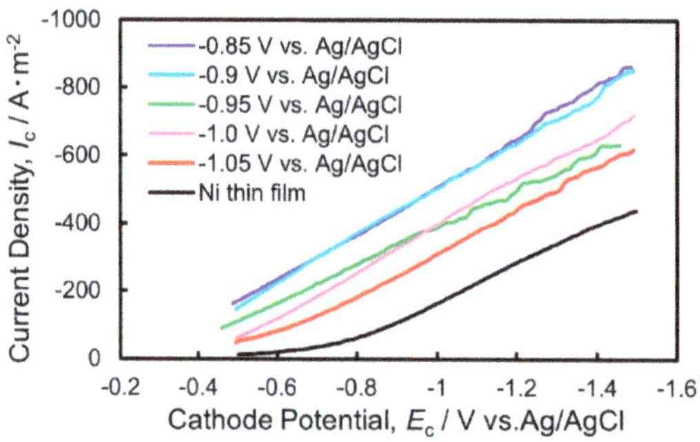

Figure 9. Cathodic polarization curves for hydrogen evolution on the electrodeposited Ni nanowire arrays to investigate the electrocatalytic performance.

The cathode current density for hydrogen evolution reaction on the Ni nanowire array electrode with a pore filling ratio of 23.9% was approximately −580 A/m^2 at the potential of −1.0 V vs. Ag/AgCl, whereas that on the Ni thin film electrode was around −170 A/m^2. Lee et al. discovered that they could fabricate Ni nanowire arrays of 20 μm in average diameter (D), 200 μm in length (L), and approximately 100 in aspect ratio (L/D) [4]. They revealed that the cathode current density at the potential of −1.0 V vs. Ag/AgCl was approximately −50 A/m^2. Furthermore, Nie et al. reported that they could synthesize Co-Ni nanowire arrays of 60 nm in average diameter (D), 60 μm in length (L), and approximately 1000 in aspect ratio (L/D) [7]. They found that the cathode current density at the potential of −1.0 V vs. Ag/AgCl was around −145 A/m^2.

In the present study, the cathode current density at −1.5 V was approximately −891 A/m^2, which was superior to the above-mentioned previous studies due to the increased aspect ratio of the Ni nanowire arrays. If the Ni nanowires which were obtained in the present study had an ideal array structure, the specific surface area would be around 1965 times larger than that of a Ni thin film. However, as shown in Figure 4a, the bundle structure which was caused by the aggregation between the nanowires seemed to inhibit the enhancement in the specific surface area. Further improvements in the electrocatalytic performance will be realized if we can precisely optimize the nanowire array structure, such as the diameter, length, and inter-wire distance.

4. Conclusions

Ni nanowire arrays with an ultra-high density were synthesized by utilizing a potentiostatic electrodeposition technique into AAO nanochannels with a length of 320 μm, diameter of 100 nm, and aspect ratio of 3200. The Ni nanowire arrays had a textured structure with a strong orientation of fcc-Ni [111] in the axial direction regardless of the electrodeposition potential. The uniaxial magnetization behavior was confirmed in the axial direction of the electrodeposited Ni nanowire arrays. Additionally, the magnetic performance of the Ni nanowires improved by increasing the aspect ratio. The squareness and coercivity increased up to 0.8 and 550 Oe, respectively. Furthermore, regarding the electrocatalytic reaction for hydrogen evolution on the Ni nanowire array cathode, the minimum hydrogen overvoltage decreased down to 0.1 V by applying the nanowire array electrode. The cathode current density for the catalytic reaction during hydrogen evolution increased up to around −580 A/m^2 at −1.0 V and −891 A/m^2 at −1.5 V vs. Ag/AgCl. The improvement could be derived from the enhancement in the specific surface area of the electrodeposited Ni nanowire arrays that had an ultra-large aspect ratio.

Author Contributions: Y.S., R.S. and M.H. carried out experiments, analyzed data, and wrote the manuscript. T.O. designed the study, supervised the project, and analyzed the data. All authors have read and agreed to the published version of the manuscript.

Funding: This work was financially supported by the Iwatani Naoji Foundation (the 49th Iwatani Science and Technology Research Grant, No. 11).

Data Availability Statement: Datasets generated during the current study are available from the corresponding author on reasonable request.

Acknowledgments: The authors are grateful for the fruitful discussion with the members of the Metallurgy and Metallic Materials Science Laboratory in Nagasaki University.

Conflicts of Interest: The authors declare that they have no competing interests.

References

1. Shahzad, A.; Khan, I.A.; Manzoor, A.; Kashif, M.; Ahsan, M. Synthesis of nickel nanowires (Ni-NWs) as high ferromagnetic material by electrodeposition technique. *Heliyon* **2023**, *9*, e12576. [CrossRef] [PubMed]
2. Mansouri, N.; Benbrahim-cherief, N.; Chainet, E.; Charlot, F.; Encinas, T.; Boudinar, S.; Benfedda, B.; Hamadou, L.; Kadri, A. Electrodeposition of equiatomic FeNi and FeCo nanowires: Structural and magnetic properties. *J. Magn. Magn. Mater.* **2020**, *493*, 165746. [CrossRef]
3. Javed, K.; Zhang, X.M.; Parajuli, S.; Ali, S.S.; Ahmad, N.; Shah, S.A.; Irfan, M.; Feng, J.F.; Han, X.F. Magnetization behavior of NiMnGa alloy nanowires prepared by DC electrodeposition. *J. Magn. Magn. Mater.* **2020**, *498*, 166232. [CrossRef]
4. Lee, J.K.; Yi, Y.; Lee, H.J.; Uhm, S.; Lee, J. Electrocatalytic activity of Ni nanowires prepared by galvanic electrodeposition for hydrogen evolution reaction. *Catal. Today* **2009**, *146*, 188–191. [CrossRef]
5. Ganci, F.; Patella, B.; Cannata, E.; Cusumano, V.; Aiello, G.; Sunseri, C.; Mandin, P.; Inguanta, R. Ni alloy nanowires as high efficiency electrode materials for alkaline electrolysers. *Int. J. Hydrogen Energy* **2021**, *46*, 35777–35789. [CrossRef]
6. Buccheri, B.; Ganci, F.; Patella, B.; Aiello, G.; Mandin, P.; Inguanta, R. Ni-Fe alloy nanostructured electrodes for water splitting in alkaline electrolyser. *Electrochim. Acta* **2021**, *388*, 138588. [CrossRef]
7. Nie, M.; Sun, H.; Gao, Z.D.; Li, Q.; Xue, Z.H.; Luo, J.; Liao, J.M. Co–Ni nanowires supported on porous alumina as an electrocatalyst for the hydrogen evolution reaction. *Electrochem. Commun.* **2020**, *115*, 106719. [CrossRef]
8. Shi, S.; Zhou, D.; Jiang, Y.; Cheng, F.; Sun, J.; Guo, Q.; Luo, Y.; Chen, Y.; Liu, W. Lightweight Zn-Philic 3D-Cu Scaffold for Customizable Zinc Ion Batteries. *Adv. Funct. Mater.* **2024**, 2312664. [CrossRef]
9. Chen, J.; Li, X.; Ma, B.; Zhao, X.; Chen, Y. CoP@Ni core-shell heterostructure nanowire array: A highly efficient electrocatalyst for hydrogen evolution. *J. Colloid Interface Sci.* **2023**, *637*, 354–362. [CrossRef] [PubMed]
10. Mohammadi, O.; Bahari, Y.; Ahmadi Daryakenari, A.; Jalali Koldeh, F.; Zhang, X.; Tian, Z.Q.; Shen, P.K. NiCoP nanoarchitectures: One-step controled electrodeposition and their application as efficient electrocatalysts for boosting hydrogen evolution reaction. *Int. J. Hydrogen Energy* **2022**, *47*, 34943–34954. [CrossRef]
11. Wang, F.; Feng, X.X.; Wang, N.; Guan, H.X.; Bian, S.K.; Hao, X.F.; Chen, Y. In-situ grown nickel-cobalt bimetallic nanowire arrays for efficient hydrogen evolution reaction. *Colloid. Surf. A Physicochem. Eng. Asp.* **2021**, *615*, 126205. [CrossRef]
12. Gao, X.; Lu, K.; Chen, J.; Min, J.; Zhu, D.; Tan, M. NiCoP–CoP heterostructural nanowires grown on hierarchical Ni foam as a novel electrocatalyst for efficient hydrogen evolution reaction. *Int. J. Hydrogen Energy* **2021**, *46*, 23205–23213. [CrossRef]
13. Wang, S.; Lu, A.; Zhong, C. Hydrogen production from water electrolysis: Role of catalysts. *Nano Conver.* **2021**, *8*, 4. [CrossRef] [PubMed]
14. Nguyen Vien, G.; Rioual, S.; Gloaguen, F.; Rouvellou, B.; Lescop, B. Study of the magnetization behavior of ferromagnetic nanowire array: Existence of growth defects revealed by micromagnetic simulations. *J. Magn. Magn. Mater.* **2016**, *401*, 378–385. [CrossRef]
15. Lavín, R.; Denardin, J.C.; Cortés, A.; Gómez, H.; Cornejo, M.; González, G. Magnetic properties of cobalt nanowire arrays. *Molec. Cryst. Liq. Cryst.* **2010**, *521*, 293–300. [CrossRef]
16. Dong, X.; Qi, M.; Tong, Y.; Ye, F. Solvothermal synthesis of single-crystalline hexagonal cobalt nanofibers with high coercivity. *Mater. Lett.* **2014**, *128*, 39–41. [CrossRef]
17. Volobuev, V.V.; Groiss, H.; Halilovic, A.; Steiner, H.; Khiar, A.; Hesser, G.; Springholz, G. Nucleation and formation of Au-catalyzed ZnTe nanowires on (001) GaAs by MBE: From planar to out-of-plane growth. *J. Cryst. Growth* **2017**, *477*, 118–122. [CrossRef]
18. Kamimura, H.; Hayashida, M.; Ohgai, T. CPP-GMR Performance of Electrochemically Synthesized Co/Cu Multilayered Nanowire Arrays with Extremely Large Aspect Ratio. *Nanomaterials* **2020**, *10*, 5. [CrossRef] [PubMed]
19. Shakya, P.; Cox, B.; Davis, D. Giant Magnetoresistance and Coercivity of electrodeposited multilayered FeCoNi/Cu and CrFeCoNi/Cu. *J. Magn. Magn. Mater.* **2012**, *324*, 453–459. [CrossRef]
20. Sanchez-Barriga, J.; Lucas, M.; Rivero, G.; Marin, P.; Hernando, A. Magnetoelectrolysis of Co nanowire arrays grown in a track-etched polycarbonate membrane. *J. Magn. Magn. Mater.* **2007**, *312*, 99–106. [CrossRef]
21. Nasirpouri, F.; Southern, P.; Ghorbani, M.; Irajizad, A.; Schwarzacher, W. GMR in multilayered nanowires electrodeposited in track-etched polyester and polycarbonate membranes. *J. Magn. Magn. Mater.* **2007**, *308*, 35–39. [CrossRef]
22. Cherevko, S.; Fu, J.; Kulyk, N.; Cho, S.M.; Haam, S.; Chung, C.H. Electrodeposition Mechanism of Palladium Nanotube and Nanowire Arrays. *J. Nanosci. Nanotech.* **2009**, *9*, 3154–3159. [CrossRef] [PubMed]

23. Sato, A.; Pennec, Y.; Shingne, N.; Thurn-Albrecht, T.; Knoll, W.; Steinhart, M.; DjafariRouhani, B.; Fytas, G. Tuning and switching the hypersonic phononic properties of elastic impedance contrast nanocomposites. *ACS Nano* **2010**, *4*, 3471–3481. [CrossRef] [PubMed]
24. Neetzel, C.; Ohgai, T.; Yanai, T.; Nakano, M.; Fukunaga, H. Uniaxial magnetization performance of textured Fe nanowire arrays electrodeposited by a pulsed potential deposition technique. *Nanoscale Res. Lett.* **2017**, *12*, 598. [CrossRef] [PubMed]
25. Nykiel, A.; Ledwig, P.; Pawlik, P.; Ghanbaja, J.; Cempura, G.; Kruk, A.; Walcarius, A.; Kac, M. The influence of electrodeposition potential on the chemical composition, structure and magnetic properties of FeCoNi nanowires. *J. Alloys Compd.* **2024**, *982*, 173709. [CrossRef]
26. Ohgai, T.; Mizumoto, M.; Nomura, S.; Kagawa, A. Electrochemical fabrication of metallic nanowires and metal oxide nanopores. *Mater. Manufact. Process.* **2007**, *22*, 440–443. [CrossRef]
27. Zhang, H.; Jia, W.; Sun, H.; Guo, L.; Sun, J. Growth mechanism and magnetic properties of Co nanowire arrays by AC electrodeposition. *J. Magn. Magn. Mater.* **2018**, *468*, 188–192. [CrossRef]
28. Thiem, L.V.; Tu, L.T.; Phan, M. Magnetization Reversal and Magnetic Anisotropy in Ordered CoNiP Nanowire Arrays: Effects of Wire Diameter. *Sensors* **2015**, *15*, 5687–5696. [CrossRef] [PubMed]
29. Vazquez, M.; Hernandez-Velez, M.; Pirota, K.; Asenjo, A.; Navas, D.; Velazquez, J.; Vargas, P.; Ramos, C. Arrays of Ni nanowires in alumina membranes: Magnetic properties and spatial ordering. *Euro. Phys. J. B* **2004**, *4060*, 489–497. [CrossRef]
30. Escrig, J.; Lavín, R.; Palma, J.L.; Denardin, J.C.; Altbir, D.; Cortés, A.; Gómez, H. Geometry, dependence of coercivity in Ni nanowire arrays. *Nanotechnology* **2008**, *19*, 075713. [CrossRef]
31. Wang, Y.T.; Yang, Z.; Wu, Q.; Liu, W.Q.; Li, Y.Q.; Zhang, H.G.; Ma, X.Y.; Cong, L.Y.; Wang, H.; Zhang, D.T.; et al. Effect of stacking faults on magnetic properties and magnetization reversal in Co nanowires. *Mater. Character.* **2022**, *187*, 111861. [CrossRef]
32. Li, H.; Wu, Q.; Yue, M.; Peng, Y.; Li, Y.; Liang, J.; Wang, D.; Zhang, J. Magnetization reversal in cobalt nanowires with combined magneto-crystalline and shape anisotropies. *J. Magn. Magn. Mater.* **2019**, *481*, 104–110. [CrossRef]
33. Gandha, K.; Mohapatra, J.; Liu, J.P. Coherent magnetization reversal and high magnetic coercivity in Co nanowire assemblies. *J. Magn. Magn. Mater.* **2017**, *438*, 41–45. [CrossRef]
34. Tishkevich, D.; Vorobjova, A.; Shimanovich, D.; Kaniukov, E.; Kozlovskiy, A.; Zdorovets, M.; Vinnik, D.; Turutin, A.; Kubasov, I.; Kislyuk, A.; et al. Magnetic Properties of the Densely Packed Ultra-Long Ni Nanowires Encapsulated in Alumina Membrane. *Nanomaterials* **2021**, *11*, 1775. [CrossRef] [PubMed]
35. Saeki, R.; Mizoguchi, S.; Kamimura, H.; Hayashida, M.; Ohgai, T. CPP-GMR performance of electrodeposited metallic multilayered nanowires with a wide range of aspect ratios. *J. Magn. Magn. Mater.* **2021**, *529*, 167849. [CrossRef]
36. Neetzel, C.; Kamimura, H.; Hayashida, M.; Ohgai, T. Uniaxial magnetization reversal process in electrodeposited high-density iron nanowire arrays with ultra-large aspect ratio. *Result. Phys.* **2019**, *15*, 102653. [CrossRef]
37. Yoshimoto, M.; Morizono, Y.; Tsurekawa, S.; Baba, T. Anodizing of aluminum in sulfuric acid and oxalic acid solutions with percarboxylic acid-based additive. *J. Ceram. Soc. Jpn.* **2012**, *120*, 276–279. [CrossRef]
38. Bockris, J. O'.M.; Hideaki, K. Analysis of galvanostatic transients and application to the iron electrode reaction. *J. Electrochem. Soc.* **1961**, *108*, 676–681. [CrossRef]
39. Munford, M.L.; Seligman, L.; Sartorelli, M.L.; Voltolini, E.; Martins, L.F.O.; Schwarzacher, W.; Pasa, A.A. Electrodeposition of magnetic thin films of cobalt on silicon. *J. Magn. Magn. Mater.* **2001**, *226–230*, 1613–1615. [CrossRef]
40. Yanai, T.; Shiraishi, K.; Akiyoshi, T.; Azuma, K.; Watanabe, Y.; Ohgai, T.; Morimura, T.; Nakano, M.; Fukunaga, H. Electroplated Fe-Co-Ni films prepared from deep-eutectic-solvent-based plating baths. *AIP Adv.* **2016**, *6*, 055917. [CrossRef]
41. Duan, J.; Liu, J.; Cornelius, T.W.; Yao, H.; Mo, D.; Chen, Y. Magnetic and optical properties of cobalt nanowires fabricated in polycarbonate ion-track templates. *NIM-B* **2009**, *267*, 2567–2570. [CrossRef]
42. Harri, G.B. X. Quantitative measurement of preferred orientation in rolled uranium bars. *Phil. Mag.* **1952**, *43*, 113–123. [CrossRef]
43. Pangarov, N.A. The crystal orientation of electrodeposited metals. *Electrochim. Acta* **1962**, *7*, 139–146. [CrossRef]
44. Pangarov, N.A. On the crystal orientation of electrodeposited metals. *Electrochim. Acta* **1964**, *9*, 721–726. [CrossRef]
45. Pangarov, N.A.; Vitkova, S.D. Preferred orientation of electrodeposited cobalt crystallites. *Electrochim. Acta* **1966**, *11*, 1733–1745. [CrossRef]
46. Yanai, T.; Shiraishi, K.; Watanabe, Y.; Nakano, M.; Ohgai, T.; Suzuki, K.; Fukunaga, H. Electroplated Fe-Ni films prepared from deep eutectic solvents. *IEEE Trans. Magn.* **2014**, *50*, 2008404. [CrossRef]
47. Fan, S.; Zhou, C.; Xu, H.; Xu, J.; Wen, H.M.; Xiao, J.Q.; Hu, J. A novel strategy to improve giant magnetoresistance effect of Co/Cu multilayered nanowires arrays. *J. Alloys Compd.* **2022**, *910*, 164729. [CrossRef]

nanomaterials

MDPI

Article

Cost-Effective Bimetallic Catalysts for Green H_2 Production in Anion Exchange Membrane Water Electrolyzers

Sabrina Campagna Zignani [1], Marta Fazio [1], Mariarosaria Pascale [1], Chiara Alessandrello [2], Claudia Triolo [2,3], Maria Grazia Musolino [2,3,*] and Saveria Santangelo [2,3]

1 Institute of Advanced Technologies for Energy (ITAE) of the National Research Council (CNR), 98126 Messina, Italy; sabrina.campagnazignani@cnr.it (S.C.Z.); fazio@itae.cnr.it (M.F.); pascale@itae.cnr.it (M.P.)

2 Department of Civil, Energy, Environmental and Materials Engineering (DICEAM), Mediterranean University of Reggio Calabria, 89122 Reggio Calabria, Italy; c.alessandrello@unirc.it (C.A.); claudia.triolo@unirc.it (C.T.); saveria.santangelo@unirc.it (S.S.)

3 National Reference Center for Electrochemical Energy Storage (GISEL), National Interuniversity Consortium for the Science and Technology of Materials (INSTM), 50121 Florence, Italy

* Correspondence: mariagrazia.musolino@unirc.it

Abstract

Green hydrogen production from water electrolysis (WE) is one of the most promising technologies to realize a decarbonized future and efficiently utilize intermittent renewable energy. Among the various WE technologies, the emerging anion exchange membrane (AEMWE) technology shows the greatest potential for producing green hydrogen at a competitive price. To achieve this goal, simple methods for the large-scale synthesis of efficient and low-cost electrocatalysts are needed. This paper proposes a very simple and scalable process for the synthesis of nanostructured NiCo- and NiFe-based electrode materials for a zero-gap AEMWE full cell. For the preparation of the cell anode, oxides with different Ni molar fractions (0.50 or 0.85) are synthesized by the sol–gel method, followed by calcination in air at different temperatures (400 or 800 °C). To fabricate the cell cathode, the oxides are reduced in a H_2/Ar atmosphere. Electrochemical testing reveals that phase purity and average crystal size significantly influence cell performance. Highly pure and finely grained electrocatalysts yield higher current densities at lower overpotentials. The best performing membrane electrode assembly exhibits a current density of 1 A cm^{-2} at 2.15 V during a steady-state 150 h long stability test with 1 M KOH recirculating through the cell, the lowest series resistance at any cell potential (1.8 or 2.0 V), and the highest current density at the cut-off voltage (2.2 V) both at the beginning (1 A cm^{-2}) and end of tests (1.78 A cm^{-2}). The presented results pave the way to obtain, via simple and scalable techniques, cost-effective catalysts for the production of green hydrogen aimed at a wider market penetration by AEMWE.

Keywords: bimetallic electrocatalysts; green hydrogen; nanostructured electrodes; anion exchange membrane water electrolysis; alkaline electrolyzer

1. Introduction

In sustainable energy societies, the development and use of clean, renewable and environmentally friendly energy is a necessary solution to address global energy demand and the rapid depletion of fossil fuels, as well as to solve the associated environmental pollution issues, such as environmental degradation and climate change [1–3]. Thanks to its pollution-free with only water and heat as combustion products, the high abundance and high energy density, hydrogen as an energy carrier is widely regarded as the most promising alternative to fossil fuels in the 21st century and will play a key role in the energy transition to achieve the goal of net zero CO_2 emissions [4–6]. Among the various sustainable technologies for H_2 production, electrochemical water splitting is one of the most efficient, feasible and clean options for generating high-purity hydrogen on a large scale, especially when it is powered by renewable energy sources [7–9].

Alkaline water electrolysis (AWE) and proton exchange membrane water electrolysis (PEMWE) are the most widely used technologies for low-temperature hydrogen production [7,10]. AWE, which works with a highly concentrated alkaline liquid electrolyte (typically an aqueous solution of 20–30 wt% KOH), a cheaper and porous diaphragm separator, and non-noble metals as electrocatalysts, represents a cost-effective and industry-proven technology [11]. However, AWE suffers from some drawbacks, such as handling a highly caustic electrolyte and low current densities. In addition, the diaphragm, failing to completely prevent the cross-over of the produced gases from one half-cell to the other, leads to lower energy efficiency and safety issues [10,12]. PEMWE technology, which uses an ultrathin polymer membrane as a solid electrolyte with high proton conductivity and a pure water feed, features a compact device design, higher current density, higher efficiency, and higher hydrogen purity than AWE [13]. Nonetheless, the presence of a harsh and acidic environment, due to the proton exchange membrane, requires the use of specific and expensive corrosion-resistant materials for electrolyzer components. Platinum group metals (PGMs)-based catalysts, like Pt for the cathode and IrO_2/RuO_2 for the anode, Nafion-type membrane, titanium-based current collectors, and bipolar plates are commonly used [14,15]. Therefore, high investment costs continue to be a major bottleneck for the large-scale application of PEMWE technology.

Recently, anion exchange membrane water electrolysis (AEMWE) has emerged as an appealing process for green hydrogen production and a potential solution to the shortcomings of both AWE and PEMWE technologies [10,16–18]. AEMWE combines the cost-effectiveness of AWE with the advantages of PEMWE, such as hydrogen purity and high current density. Similar to PEMWE, this technology employs a thin anion exchange membrane (AEM) as the solid polymer electrolyte and a zero-gap cell configuration, enabling a compact cell design with lower ohmic resistance, higher operating current densities, and, by efficiently separating hydrogen, high-purity hydrogen production. Furthermore, since the AEM conducts hydroxide ions, an alkaline working environment is present as in the AWE, which allows the use of non-precious transition metals and their derivatives as electrocatalysts and inexpensive materials for membranes and bipolar plates, improving economic feasibility [10,16–21].

Overall electrochemical water splitting process occurs through two half-cell reactions: hydrogen evolution reaction (HER) at the cathode and oxygen evolution reaction (OER) at the anode. Under standard conditions, a theoretical thermodynamic potential of 1.23 V is required to drive the water splitting process, corresponding to an energy consumption of $\Delta G = 237.1 \text{ kJ mol}^{-1}$ [19,22]. However, a higher cell voltage (overpotential) is necessary to overcome the intrinsic energy barriers due to the sluggish kinetics of the HER and OER reactions, especially for the OER involving a 4-electron transfer and more intermediates, and the ohmic resistance of the electrolyte and cell components [10,18,23,24]. Overpotentials result in high energy consumption and low energy conversion efficiency. Therefore, the development of highly active, low-cost, sustainable, and durable catalysts is crucial to simultaneously accelerate the kinetics of HER and OER and efficiently reduce the overpotentials, improving the performance of electrochemical water splitting.

Over the past decade, first-row transition metal-based compounds and, in particular, Ni-based materials, including alloys [25,26], oxides [27,28], layered double hydroxide [29,30], chalcogenides [31,32], phosphides [33,34], and nitrides [29,35], have been developed as promising OER and HER catalysts for the AEMWE, due to their stability and corrosion resistance in an alkaline environment. However, most of the reported works are limited to half-cell studies. Furthermore, catalysts with outstanding activity towards OER often show poor HER activity, due to the different mechanisms. As a result, Pt-based benchmark catalysts continue to be the primary cathodic material in an AEMWE single-cell configuration to achieve low overpotentials and high current densities. Thus, the design of effective and stable non-PGM bifunctional catalysts for both OER and HER reactions in an alkaline environment remains an urgent challenge. The use of the same material as a catalyst at both the anode and the cathode leads to a simplification of the electrolysis process and a significant reduction in the overall cost, because, for example, different equipment and methodologies are not required to produce different electrocatalysts. Although some progress has been made, very few studies that meet the above requirements and show long-term operational lifetime are reported on the overall water splitting in AEMWE [18,22].

Considering the above context, the focus of this work is to develop electrocatalysts through a very simple and scalable synthesis method and to study their electrocatalytic activity and long-term stability towards water electrolysis in an AEMWE single-cell configuration. It is widely recognized that the combination of non-precious metals, such as nickel/cobalt and nickel/iron, in oxides/hydroxides/oxyhydroxides leads to sustainable electrocatalysts with interesting OER properties. This is often understood as the effect of the synergistic interaction between the two metals, potentially resulting in enhanced conductivity and the formation of more active sites. Furthermore, the structure and catalytic activity of bimetallic catalysts can be influenced by the ratio of the two metals [36,37]. Here, nanostructured NiCo- and NiFe-oxides, with different concentrations of the two metals (50:50 or 85:15), are produced by the sol–gel method followed by calcination in air at two different temperatures (400 or 800 °C) to investigate the effect of crystal size and crystallization degree. The as-prepared and H_2/Ar-reduced binary oxides are used as electrocatalysts to fabricate the anode and cathode, respectively. The electrodes are then assembled in a zero-gap AEMWE full cell, in combination with a commercial polymeric anion exchange membrane (Fumatech® FAA3-50) and a commercial ionomer (ION FAA-3-SOLUTION). Reference monometallic (nickel) oxide and hydroxide are also prepared and tested. Electrocatalysts are thoroughly characterized by complementary analysis techniques to investigate the structural factors playing a key role in their electrocatalytic performance.

2. Materials and Methods

2.1. Synthesis of OER Electrocatalysts

Nanostructured bimetallic oxides to be used as OER electrocatalysts at the cell anode were prepared via the sol–gel (SG) method and subsequent calcination, following the procedure described in detail in a previous paper [38]. Nickel (II) acetate tetrahydrate (purity > 98%, CAS No. 6018-89-9, Merck), cobalt (II) acetate tetrahydrate (purity > 98%, CAS No. 6147-53-1, Merck), and iron acetate (purity > 95%, CAS No. 3094-87-9, Merck) were used as Ni, Co, and Fe sources, respectively. Stoichiometric amounts of these salts (Table 1) were dissolved in 60 g of water, one at a time (Figure S1a). After magnetic stirring at 250 rpm for 30 min, 6 g of monohydrate citric acid (purity 98%; CAS No. 5949-29-1, Merck) was added as a complexing agent, and the resulting mixture was further stirred at 350 rpm for 2 h at 70 °C to form a gel. The as-obtained gel was dried at 80 °C overnight and, subsequently, calcined in static air in a muffle furnace (Figure S1b). Temperature was increased at a rate of 10 °C min^{-1} and kept constant for 2 h at the selected value (T_C, Table 1). Calcination was followed by rapid cooling down to room temperature (RT) out of the furnace to generate defects on the oxide surface [38–40] and enhance porosity and oxygen deficiency [41].

Table 1. Codes, composition, and calcination temperature (T_C) of the nanostructured oxides prepared via the sol–gel (SG) method.

Code	Metal Molar Concentrations			T_C/°C
	Ni	Co	Fe	
Ni100	1.00			400
Ni85Co15_400	0.85	0.15		400
Ni85Co15_800	0.85	0.15		800
Ni50Co50_400	0.50	0.50		400
Ni50Co50_800	0.50	0.50		800
Ni85Fe15_400	0.85		0.15	400
Ni85Fe15_800	0.85		0.15	800

A reference monometallic (nickel) oxide was further prepared via the same procedure (Table 1). In addition, a reference nickel hydroxide sample (coded as NiH100) was synthesized from nickel (II) nitrate hexahydrate (purity > 99%, CAS No. 13138-45-9, Aldrich) via the co-precipitation (CP) method. Nickel nitrate hexahydrate was dissolved in ultrapure distilled water at 60 °C to form a suspension that was neutralized at pH 9 with a 1 M NaOH solution. The suspension was stirred and kept at 60 °C and pH 9 for 4 h to promote the precipitation of the compound. The precipitate was then filtered, washed with heated ultrapure water, and dried at 80 °C overnight. The scheme of electrocatalyst preparation is provided in Figure S2.

2.2. *Synthesis of HER Electrocatalysts*

A portion of the as-prepared oxides was utilized to produce HER electrocatalysts for the cell cathode (Figure S3a). Oxide powders were reduced in 5% H_2/Ar atmosphere for 30 min (Figure S3b) to form reduced bimetallic compounds (Figure S3c). The reduction temperatures (T_R) of each material, reported in Table S1, were determined by means of the temperature programmed reduction (TPR) analysis (Figure S4), carried out on an AMI-300 instrument.

2.3. *Electrocatalysts Characterization*

All electrocatalysts were characterized through scanning electron microscopy (SEM) and energy-dispersive X-ray (EDX) analysis using an SEM-FEG-UHR microscope (Thermo Fisher) equipped with a FIB (focused ion beam) column and STEM (scanning transmission electron microscopy) detectors. The instrument was operated in the range 2–15 kV, and the EDX probe was used to determine the bulk elemental composition of the samples.

The oxide phases formed under calcination in OER catalysts and the metallic species obtained by reduction in HER catalysts were identified by X-ray diffraction (XRD). The analysis was carried out with a Bruker D2 diffractometer using Ni β-filtered Cu-K_α radiation source (λ = 0.1541 nm). As usual [42], to determine the phase composition of the electrocatalysts, diffractograms were analyzed by the Rietveld method using Maud 2.992 software and the isotropic model for all samples except for NiH100. For the latter sample, the Popa model was used [43] in order to more accurately describe the anisotropic behavior of the crystallites, which is well documented in the literature [44,45].

The spatial homogeneity of the oxides was assessed by measuring Raman scattering from various random positions on each specimen. For this purpose, an NTEGRA—Spectra

SPM NT-MDT confocal microscope coupled to a solid-state laser operating at 532 nm was used. Measurements were carried out in air at RT by using a low laser power (250 μW at the sample surface) to prevent local heating. The scattered light from the sample was collected by a 100X Mitutoyo objective (NA = 0.75), dispersed by an 1800 lines mm^{-1} grating, and detected by a cooled ANDOR iDus CCD Camera. Since the surface area of the region probed by micro-Raman spectroscopy (MRS, <0.6 μm^2) is smaller than that probed by XRD, to assess spatial homogeneity of the oxides, spectra were recorded from random positions on each specimen and then averaged to infer reliable information on the entire sample.

The surface chemical composition of the samples and chemical environment were evaluated by X-ray photoelectron spectroscopy (XPS). Spectra were recorded using a Physical Electronics GMBH PHI 5800-01 spectrometer (Physical Electronics GmbH, Munich, Germany), equipped with a monochromatic Al-K_α source (1486.6 eV) with a power beam of 300 W. XPS data were interpreted using the online library of oxidation states implemented in PHI MultiPak 6.1 software (Chanhassen, MN, USA, 1999) and the PHI Handbook of X-ray photoelectron spectroscopy.

2.4. Anionic Electrolyte

A commercial Fumatech® FAA3-50 membrane was used as a solid polymer electrolyte in AEM electrolysis. The selected thickness of the membrane (50 μm) was chosen from a compromise between low cross-over, low area-specific resistance, and suitable mechanical stability requirements. FAA3-50 membranes contain ammonium functional groups, allowing the exchange of hydroxyl ions, characterized by a lower mobility compared to protons, from the cathode compartment to the anode, where the OER takes place. By doing so, the recombination of the reaction products can be avoided while ions can percolate in the device as needed to complete the electrochemical process. Before use, the membrane needs to be activated by treating it with a 3 M NaCl solution for 72 h in order to obtain the membrane in Cl^- form. This step is necessary to facilitate the exchange between Cl^- and OH^- ions before cell assembly. A commercial ionomer (ION FAA-3-SOLUTION) from Fumatech® was utilized as a binding agent during ink preparation in order to improve the adhesion of the electrocatalyst powder to the surface support.

2.5. Membrane Electrode Assembly and Electrochemical Studies

To prepare the membrane electrode assemblies (MEAs), the first step was to mix together the anionic ionomer, electrocatalyst powder, and solvent to create the anode and cathode inks for spray deposition onto specific supports. The ionomer, used to prepare the inks, was a commercial ION FAA-3-SOLUTION (10 %wt) from Fumatech®. Spray coating was used to deposit the catalysts on the support layers after dispersing them in ethanol and sonicating them for 30 min (Figure S5). To facilitate ethanol evaporation, a heated plate was used. A correct distance was maintained between the airbrush and the support to avoid ink penetration. Figure S6 shows a typical anode and cathode, Fumatech® membrane, and MEA with a 5 cm^2 active area used for the electrochemical test.

The anode inks, based on catalysts in their oxidized form (Table 1), were deposited on a nickel felt (Bekaert). Gas diffusion layers (GDLs) based on carbon paper (39BB, SIGRACET) were used as a support to deposit the cathodic inks, based on catalysts in their reduced form (Table S1). The mass loads were about 2.5 mg cm^{-2} and 3 mg cm^{-2} for the anode and cathode electrocatalysts, respectively.

The anode and cathode compartments were separated by a commercial Fumatech® anionic exchange membrane (50 μm thickness) in the OH^- form. Before the assembly, the membrane and the electrodes, containing the ionomer, were exchanged for 24 h with a solution containing hydroxide ions in order to activate the components.

Figure S7 displays the operation scheme for the AEM electrolysis cell and its functional components. A cold assembly procedure was followed to prepare the MEAs to avoid any undesired degradation of the membrane, as observed during the conventional hot-pressing lamination process. The MEAs were assembled in a single-cell housing made of nickel plates characterized by a serpentine flow field channel that matches the active area of 5 cm^2. Teflon® gaskets were used to seal the cell and avoid any leakage of the electrolyte solution. Cell compression was 2.5 N m per tie rod, and this tightening for the cell can ensure that the electrodes and the membrane are in direct contact like a real sandwich. The single cell was tested at 50 °C and under atmospheric pressure. During all experiments, the anode compartment was fed with 1 M KOH solution at a flow rate of 5 mL min^{-1} using a peristaltic pump. The cell was maintained at the same temperature using heating mats applied to the external plates of the cell.

2.6. Electrochemical Studies

The system was characterized electrochemically to study the performance of the MEAs in terms of reaction rates at different cell potentials. To investigate the electrochemical behavior of each MEA, galvanostatic polarization curves (cell voltage versus current density) were recorded, and galvanostatic durability tests (cell voltage versus time) and electrochemical impedance spectroscopy (EIS) analyses were performed. Polarization curves and chrono-potentiometric curves were carried out with a Keithley power supply system (Tektronic), a PGSTAT Autolab 302 Potentiostat/Galvanostat equipped with a booster of 20 A (Metrohm), and a Frequency Response Analyzer (FRA) was used for EIS analyses. EIS measurements were performed under potentiostatic control in a frequency range between 1 MHz and 10 mHz by frequency sweeping in a single sine mode and acquiring ten points per decade. The amplitude of the sinusoidal excitation signal was 0.01 V r.m.s. The series resistance was determined from the high frequency intercept on the real axis in the Nyquist plot, while the polarization resistance was estimated as the difference between the extrapolated low frequency intercept and the high frequency intercept on the real axis [30].

3. Results and Discussion

3.1. OER Electrocatalysts

3.1.1. Morphology

The morphology and composition of all electrocatalysts were investigated by SEM and SEM/EDX. Figure 1 shows some representative SEM images of the OER electrocatalysts investigated. Lower magnification micrographs are shown in Figure S8a–c. The results of elemental analysis by SEM/EDX are reported in Figure S8d–f. Regardless of the synthesis method, large agglomerates of small particles are formed, as proven by higher magnification images (Figure 1g). In the case of the sample (NiH100) prepared via the CP method, the particle aggregates look more compact.

The results of SEM/EDX analysis demonstrate that the organic components of the metal precursors are completely removed during calcination, with the formation of inorganic oxides. The identification of their phase(s) was carried out via XRD analysis.

3.1.2. Phase(s) and Average Crystallite Size of the Oxides

Figure S9 displays the diffractograms of the reference monometallic samples; the XRD patterns of bimetallic oxides are shown in Figure 2a–c. All diffraction peaks detected in the reference NiH100 sample (top of Figure S9b) can be indexed to the β-phase of nickel hydroxide [46,47]. This evidences the formation of a single phase. Similarly, only the reflections from the (111), (200), (220), (311), and (222) crystallographic planes of cubic NiO with rock-salt (RS) structure at ca. 37.30°, 43.34°, 62.89°, 75.45°, and 79.40°

2θ-angles, respectively, are detected in the monometallic oxide (Ni100) prepared by SG (top of Figure S9a) [48–50].

Figure 1. Morphology of the oxides as resulting from SEM analysis. The shown micrographs refer to samples (**a**,**e**) NiH100, (**b**) Ni100, (**c**,**f**,**g**) Ni85Co15_400, and (**d**,**h**) Ni85Fe15_400.

In the case of bimetallic oxides (Figure 2a–c), XRD analysis reveals that the phase purity of the oxide is influenced by the metal pair and their molar fractions and, in some cases, also by the temperature at which the oxidative heat treatment is operated. A single RS phase forms at any T_C in Ni85Co15 oxides (Figure 2a) [48–50]. Since the radius of the Co^{2+} ion is similar to that of the Ni^{2+} ion (58 pm against 55 pm for tetrahedral coordination, respectively [51]), the substitution of a small amount (0.15 molar fraction) of Ni with Co in the cubic RS lattice easily takes place and results a uniform dispersion of Co into the NiO matrix [52] with no secondary phase formation.

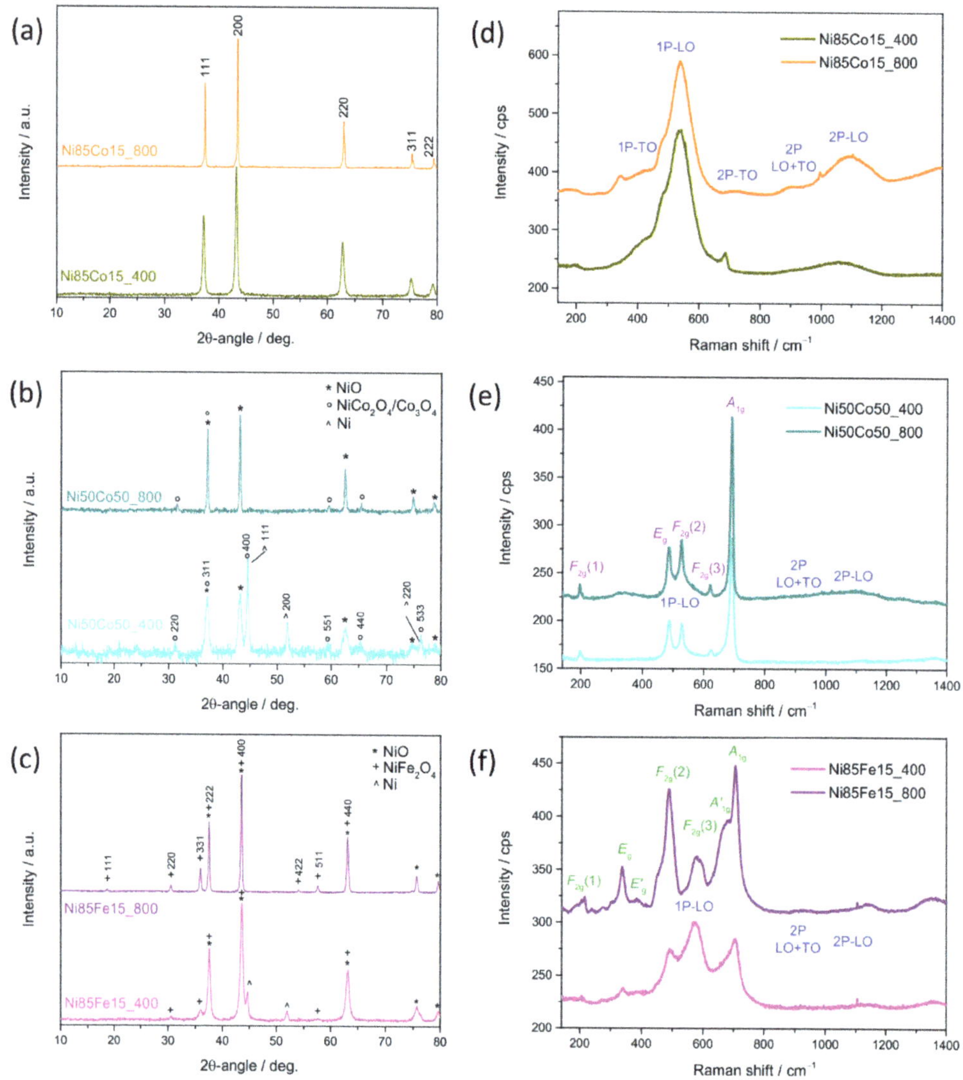

Figure 2. Results of (**a**–**c**) XRD and (**d**–**f**) MRS analyses on bimetallic oxides. The shown diffractograms refer to (**a**) Ni85Co15, (**b**) Ni50Co50, and (**c**) Ni85Fe15 electrocatalysts. Micro-Raman spectra refer to (**d**) Ni85Co15, (**e**) Ni50Co50, and (**f**) Ni85Fe15 electrocatalysts.

As the molar fraction of Co rises to 0.50, additional reflections appear in the XRD patterns of NiCo-based oxides (Figure 2b). This finding suggests that the segregation of a portion of Co/Ni in different phase(s) takes place. At higher T_C (Ni50Co50_800), the reflections from the (220), (331), (511), and (440) crystallographic planes of a phase with spinel (SP) structure are detected at ca. 31.38°, 37.60°, 59.55°, and 65.37° 2θ-angles, along with the signals arising from the RS lattice (top of Figure 2b). Thus, the increase in the Co molar fraction up to 0.50 promotes the formation of a biphasic oxide, in agreement with other literature reports [53,54]. Besides NiO, cobalt (II, III) oxide (Co_3O_4) and/or nickel cobaltite ($NiCo_2O_4$) formed. In Ni50Co50_400, the reflections from Ni/α-Co metals with cubic structure are further clearly visible in the diffraction pattern (bottom of

Figure 2b) [49,50,55], revealing that the oxidation upon thermal treatment is incomplete, and as a result, a three-phase material is formed.

NiFe-based oxides (Figure 2c) behave similarly to the equimolar NiCo-based oxides. In the diffractogram of Ni85Fe15_800, in addition to the reflections from the cubic RS NiO lattice, those originating from the crystallographic planes of a SP-structured phase appear (top of Figure 2c), indicating the formation of a biphasic oxide, in line with literature reports [56]. Nickel ferrite ($NiFe_2O_4$) forms together with cubic NiO. At lower T_C (Ni85Fe15_400), the reflections from Ni/Co metals are also visible (bottom of Figure 2c) [49,50,55].

Additional information on the material structure and composition was inferred by carrying out Rietveld refinements from XRD data (Figure S10). The main results obtained are summarized in Table 2 and Figure S11; further details are reported in Table S2. The formation of a pure single-phase RS structure is confirmed in the case of monometallic Ni100 oxide and bimetallic Ni85Co15 oxides, regardless of their calcination conditions. In these materials, the average crystallite size (d) increases in the order Ni100 < Ni85Co15_400 << Ni85Co15_800 (Table 2).

Table 2. Main results of Rietveld analysis on the OER electrocatalysts. The relative abundance of each phase is reported, together with the mean crystallite size (d). In the case of a multiphase material, the d-value is calculated as a weighted average over the phases present. RS, CM, and SP stand for rock-salt, cubic metal/alloy, and spinel structure, respectively.

Code	Phase(s) Abundance/wt%				d/nm
	β-Ni(OH)2	RS	CM	SP	
NiH100	100.0				24
Ni100		100.0			21
Ni85Co15_400		100.0			36
Ni85Co15_800		100.0			105
Ni50Co50_400		46.8	30.6	22.6	42
Ni50Co50_800		72.2		27.8	105
Ni85Fe15_400		77.2	12.4	10.4	28
Ni85Fe15_800		81.4		18.6	55

In the remaining bimetallic oxides, two or three phases coexist, depending on T_C. RS is always the primary phase. Nonetheless, at 800 °C, its amount ranges between 72 wt% (in Ni50Co50_800) and 81 wt% (in Ni85Fe15_800); whereas, at lower T_C, it reduces to 77 wt% in Ni85Fe15_400 and drops down to 47 wt% in Ni50Co50_400. The comparison between oxides with the same nominal composition reveals that more oxidized species are formed at higher T_C, as expected [41]. In fact, in Ni50Co50_800, a larger amount of $Co_3O_4/NiCo_2O_4$ is present compared to Ni50Co50_400 (27.8 against 22.6 wt%); analogously, in Ni85Fe15_800, $NiFe_2O_4$ is more abundant than in Ni85Fe15_400 (18.6 against 10.4 wt%). Furthermore, by comparing the composition of bimetallic materials calcined at 400 °C, it emerges that the metal/alloy component, absent in Ni85Co15_400, increases from 12.4 wt% in Ni85Fe15_400 up to 30.6 wt% in Ni50Co50_400. Finally, larger d-values are obtained upon calcination at 800 °C due to sintering effects (Table 2). At 400 °C, the presence of Co/a larger molar fraction of Co seems to favor the development of larger crystallites (compare d-values of Ni85Fe15_400 and Ni85Co 15_400, and of Ni85Co 15_400 and Ni50Co 50_400).

As a general behavior, some small shifts are observed when comparing the patterns of the samples with the same nominal composition calcined at different temperatures (Figure 2). Such shifts can be ascribed to different levels of microstrain affecting the lattice [42]. As shown in Table S2, for each oxide phase, the local distortions of the crystal structure and microstrain generally decrease in better crystallized samples (calcined at higher temperatures).

3.1.3. Spatial Uniformity and Crystallization Degree of the Oxides

Spatial uniformity of the OER electrocatalysts was assessed by measuring Raman scattering from random positions on each specimen (Figures S12 and S13). Except for sample Ni85Fe15_400 (Figure S13c), where even remarkable dissimilarities are observed between spectral profiles recorded at different locations, no significant differences are visible in the remaining samples, which proves the spatial uniformity of the oxide phase(s) formed.

Figure 2d–f displays the averaged micro-Raman spectra. Cubic RS structure belongs to the $Fm\bar{3}m$ space group. Two one-phonon (1P) modes and three two-phonon (2P) modes are predicted for this group [57,58]. The former modes comprise one transverse optical (TO) and one longitudinal optical (LO) mode; the latter includes the overtones of TO and LO modes and a combination band. 1P-TO and 1P-LO modes are symmetry-forbidden in a perfect cubic lattice [57,59]. Their detection (at ~410 and ~540 cm^{-1}, respectively) in Ni100 (Figure S12) and Ni85Co15 oxides (Figure 2d) is indicative of the presence of lattice distortion, defect-induced disorder, and surface effects [57,59]. Also, the 2P-TO (at ~730 cm^{-1}), 2P-TO+LO (at~905 cm^{-1}), and 2P-LO (at ~1080 cm^{-1}) are clearly visible in the higher frequency region of the spectra. These findings confirm the indications provided by XRD analysis. Moreover, in Ni85Co15_800 oxide, featured by larger-sized crystallites, the relative (to 2P-LO) intensity of the disorder-activated 1P-LO mode is weaker than in Ni85Co15_400, which indicates a reduced extent of structural disorder [57].

Spinel-structured phase belongs to the $Fd\bar{3}m$ space group. Based on the factor group analysis, five normal Raman-active vibration modes (A_{1g} + E_g + $3F_{2g}$) are predicted for this space group [60–62]. Both their positions and their relative intensities remarkably vary within the spinel family. In Ni50Co50 oxides (Figure 2e), at any T_C, the sharp A_{1g} mode at 691 cm^{-1} dominates the spectra. The weaker $F_{2g}(1)$, E_g, $F_{2g}(2)$, and $F_{2g}(3)$ modes are detected at frequencies (195, 483, 525, and 620 cm^{-1}, respectively) very close to those reported for crystalline Co_3O_4 [62–65]. In the spectrum of Ni50Co50_800, the narrow E_g and $F_{2g}(2)$ peaks originating from the SP-structured component of the oxide are superimposed to the broad band arising from the 1P-LO mode of the defective RS phase; in the high-frequency region, the very weak 2P modes of this phase are observed, in agreement with the evidences emerged from XRD analysis.

In the spectra of Ni85Fe15 oxides (Figure 2f), along with the sharp peaks associated to the normal Raman vibration modes of the SP-structured $NiFe_2O_4$ phase at 214 ($F_{2g}(1)$), 335 (E_g), 486 ($F_{2g}(2)$), 578 ($F_{2g}(3)$), and 704 cm^{-1} (A_{1g}) [60,66,67], two shoulders appear on the higher-frequency side of the E_g band (at 385 cm^{-1}) and on the lower-frequency side of the A_{1g} band (at 674 cm^{-1}). As known [66], nickel ferrite crystallizes in the inverse SP structure, with the tetrahedral 8*a* sites fully occupied by Fe^{3+} cations and the octahedral 16*d* sites occupied by both Ni^{2+} and Fe^{3+} cations. Thus, the additional inversion-induced spectral features (E'_g at 385 cm^{-1} and A'_{1g} at 674 cm^{-1}) are due to distortions in the lattice caused by a distribution of distances between Ni and Fe with oxygen [67]. In the lower-frequency spectral region, the modes of the $NiFe_2O_4$ component of the Ni85Fe15 oxides are superimposed on the weaker ones originating from the defective RS-structured component.

3.2. HER Electrocatalysts

HER electrocatalysts for the cell cathode were prepared via reduction of the as-produced oxide powders at T_R (Table S1) in 5% H_2/Ar atmosphere for 30 min. Figure S4 shows the TPR spectra. A single H_2 consumption peak dominates the spectra of reference Ni100 and NiH100 samples (Figure S4a); the peak is located at 450 and 300 °C, respectively, indicating that in NiH100, the reduction of Ni^{2+} to Ni^0 is easier. As a general result, in the bimetallic oxides calcined at higher temperatures (better crystallized) [41], the reduction of metal cations takes place at higher temperatures (compare the highest H_2 consumption peaks in Figure S4b–d). In these materials, the lower-T H_2 consumption peaks can be ascribed to the reduction of Ni, while the higher-T ones can be associated with the multi-step reduction of Co and Fe [68].

3.2.1. Morphology

SEM analysis (Figure 3 and Figure S14) reveals that the reduction process does not produces substantial changes in the morphology of the electrocatalysts prepared by the SG method (Figure 3b–d,f–i), while that prepared by via the CP method (Figure 3a,e) look less compact after the exposure to the H_2/Ar atmosphere. In any case, the macro-aggregates are still made up of small particles, as clearly visible in Figure 3f–i.

Figure 3. Morphology of the samples after reduction, as resulting from SEM analysis. The shown images refer to samples (**a**,**e**) NiH100R, (**b**,**f**) Ni100R, (**c**,**g**,**h**) Ni85Co15_400R, and (**d**,**i**) Ni85Fe15_400R.

Elemental analysis by SEM/EDX shows total or substantial removal of oxygen (compare Figures S8 and S15), confirming the successful reduction of materials in the oxidized form.

3.2.2. Crystalline Phase(s) and Average Crystallite Size

The phases formed from the pristine oxides upon reduction in H_2/Ar atmosphere at a proper temperature (Table S1) were investigated by XRD. Only the reflections at ca. 44.38°, 51.73°, and 76.26° 2θ-angles from the crystallographic planes of face-centered cubic (fcc) nickel [49,50] are detected in the two reference samples Ni100R and NiH100R (bottom of Figure S9a,b, respectively), confirming the full reduction of the pristine NiO and β-$Ni(OH)_2$ phases, respectively.

The diffraction patterns recorded on the materials obtained by reducing bimetallic oxides are shown in Figure 4. XRD analysis demonstrates that obtaining a single bimetallic phase (alloy) strongly depends not only on the composition of the pristine oxide but also on its preparation conditions. Only the reflections from the (111), (200), and (220) crystallographic planes of Ni/α-Co metals with fcc structure [50,55] are detected in Ni85Co15_800R (top of Figure 4a) and Ni50Co50_800R electrocatalysts (top of Figure 4b), indicating the formation of a pure NiCo alloy.

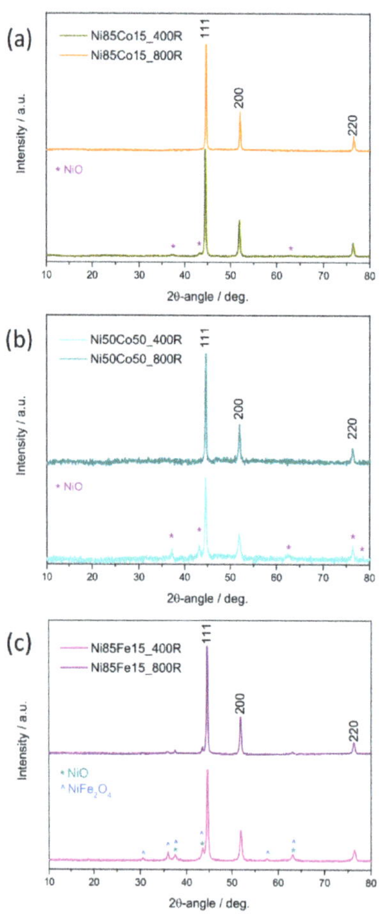

Figure 4. XRD patterns of HER electrocatalysts. The shown data refer to samples (**a**) Ni85Co15R, (**b**) Ni50Co50R, and (**c**) Ni85Fe15R.

Conversely, signals arising from the NiO component of the pristine oxides are still visible in the diffractograms of Ni85Co15_400R (bottom of Figure 4a) and Ni50Co50_400R (bottom of Figure 4b), indicating that the reduction is not complete, in agreement with the indications provided by elemental analysis by SEM/EDX. Thus, these electrocatalysts are biphasic in nature: Besides the NiCo alloy, they contain NiO as a secondary phase. Moreover, the stronger relative intensity of the NiO diffraction peaks in Ni50Co50_400R indicates the lower phase purity of this catalyst.

In Ni85Fe15_400R (bottom of Figure 4c), reflections from crystallographic planes of NiO and $NiFe_2O_4$ are detected, along with those from the fcc structure typical of Ni-rich NiFe alloy [69]. Thus, the electrocatalyst consists of three phases and has poor phase purity. A higher phase purity degree pertains to Ni50Co50_800R (top of Figure 4c), in whose diffractogram signals from the unreduced pristine oxides are extremely weak and hardly visible.

In order to get further information on the material structure and composition, Rietveld refinements from XRD data were carried out (Figure S16). Table 3 summarizes the main results obtained. Further details can be found in Table S3. Rietveld analysis confirms that the highest phase purity degree pertains to reference Ni100R and NiH100R electrocatalysts (containing Ni metal only) and to bimetallic Ni85Co15_800R and Ni50Co50_800R electrocatalysts (consisting of a pure NiCo alloy). In the former, the average size of Ni metal crystallites is around 61 nm. In Ni85Co15_800R and Ni50Co50_800R electrocatalysts, very large NiCo alloy crystallites are obtained (89 and 139 nm, respectively).

Table 3. Main results of Rietveld analysis on the HER electrocatalysts. The relative abundance of each phase is reported, together with the mean crystallite size (*d*). In the case of multiphase material, the *d*-value is calculated as a weighted average over the phases present. RS, CM, and SP stand for rock-salt, cubic metal/alloy, and spinel structure, respectively.

Sample	Phase(s) Abundance/wt%			*d*/nm
	CM	RS	SP	
NiH100R	100.0			61.3
Ni100R	100.0			61.0
Ni85Co15_400R	93.3	6.7		55.9
Ni85Co15_800R	100.0			139.4
Ni50Co50_400R	81.8	18.2		39.4
Ni50Co50_800R	100.0			88.6
Ni85Fe15_400R	71.7	18.5	9.8	38.5
Ni85Fe15_800R	91.1	2.9	6.0	59.5

In all remaining HER electrocatalysts, the reduction of the pristine oxides is not complete. The phase purity worsens in the order Ni85Co15_400R > Ni85Fe15_800R > Ni50Co50_400R > Ni85Fe15_400R, with the relative amount of residual oxide phase(s) ranging from 6.7 wt% in Ni85Co15_400R to 28.3 wt% in Ni85Fe15_400R. The average size of the crystallites is 39 nm in both Ni85Fe15_400R and N50Co50_400R and larger in the remaining impure alloys (Table 3).

3.2.3. Nanomaterial Selection

Figure S17a,b summarize the properties of the produced electrode materials in terms of phase purity and mean crystal size. Crystal size seems to be particularly critical for the anode electrocatalysts: Smaller-sized crystals, featured by larger exposed surface available for reaction and greater amounts of grain boundaries, generally exhibit a higher density of reactive surface defects [38,41,70], which is beneficial to OER [41]. Phase purity seems to play a pivotal role, too. Actually, very recently, better electrochemical performance has been reported for pure single-phase multi-component oxides [71]. Accordingly, further physicochemical characterization (XPS analysis) and electrochemical tests are performed on a restricted set of five pairs of samples, selected by choosing the best compromise between high phase purity and small average crystal size.

3.2.4. Species on the Surface of OER Electrocatalysts

The species present on the electrocatalyst surface were identified by XPS. XPS survey spectra (Figure S18) confirm the presence of nickel, oxygen, and adventitious carbon in all electrocatalysts. Cobalt is further present on the surface of Ni85Co15_400 oxide, while iron is detected on the surface of Ni85Fe15_400 and Ni85Fe15_800. High-resolution X-ray photoelectron (HRXPS) spectra of the core levels in bimetallic and reference monometallic oxides are displayed in Figure 5 and Figure S19, respectively. They provide qualitative and quantitative insights into the oxidation states of surface species in the electrocatalysts.

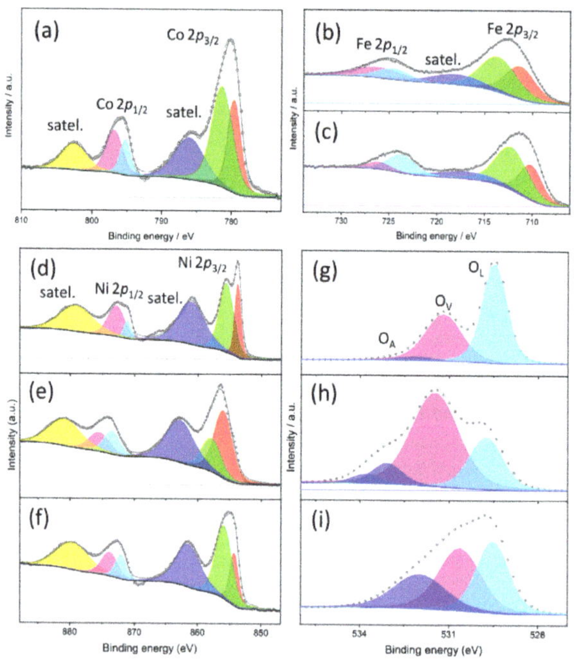

Figure 5. High-resolution XPS curves and fitting in the regions of (**a**) Co 2*p*, (**b**,**c**) Fe 2*p*, (**d**–**f**) Ni 2*p* and (**g**–**i**) O 1*s* core levels of (**a**,**d**,**g**) Ni85Co15_400, (**b**,**e**,**h**) Ni85Fe15_400, and (**c**,**f**,**i**) Ni85Fe15_800 oxides.

In Ni85Co15_400, the strong satellite peaks appearing in the spectrum of Co 2*p* core level (Figure 5a) at ca. 786 and 803 eV suggest that cobalt is mainly present as Co(II) [72]. Two different divalent cobalt species, in a 37:63 ratio, contribute to the Co $2p_{3/2}$ region, namely, Co(II) in CoO environment (at ca. 779.6 eV), and Co(II) in $Co(OH)_2$ environment (at 781.3 eV) [73,74]. Recent studies on OER electrocatalysts have reported that the presence of hydroxide species on the defective catalyst surface is beneficial for the surface restructuring process during activation [74]. In Ni85Fe15_400 and Ni85Fe15_800, two different iron species contribute to the main $2p_{3/2}$ spin-orbit component (Figure 5b,c), namely, Fe(II) in FeO environment (at ca. 711.3 eV) and Fe(III) in Fe_2O_3 environment (at 713.6 eV) [75,76], in agreement with the results of XRD and MRS analyses that indicate the co-presence of RS and SP phases. The relative amount of Fe(II) species decreases from 44% in Ni85Fe15_400 to 37% Ni85Fe15_800, while the relative amount of Fe(III) species oppositely increases from 56% to

63%, as expected due to the increase in the calcination temperature [41]. Regardless of the details, in both samples, Fe(III) species are more abundant than Fe(II) ones. Furthermore, the lack of any obvious contribution from metallic iron, in spite of the formation of a CM phase revealed by Rietveld refinements to the XRD data, is in line with other literature reports on the presence of more oxidized species on the surface of the oxides [41].

Indeed, the HRXPS spectrum in the region of Ni 2*p* core level (Figure 5d) closely resembles that of defective NiO [77]. This finding also agrees with the detection of a pure RS phase in Ni85Co15_400. The main Ni $2p_{3/2}$ spin-orbit component clearly reveals two distinct features at approximately 853.7 and 855.5 eV. They correspond to Ni(II) in NiO environment (32%) and $Ni(OH)_2$ environment (68%), respectively [74,77,78]. In monometallic Ni100 oxide (Figure S19a), the two peaks are located at slightly higher binding energies (BEs, 854.0 and 855.9 eV) and are in comparable areal ratio (34:66). A similar situation is observed in In Ni85Fe15_800 (Figure 5f) with 854.2 and 855.9 eV peaks in 29:71 areal ratio. In Ni85Fe15_400 (Figure 5e), the areal ratio inverts (66:34) and the peaks shift further towards higher BEs (856.0 and 857.9 eV), as in the case of nickel ferrite [79], again in agreement with the results of XRD and MRS analyses.

In reference to NiH100 oxide (Figure S19b), the $2p_{3/2}$ and $2p_{1/2}$ spin-orbit components (855.4 and 873.0 eV, respectively) exhibit the spin-energy separation (17.6 eV) peculiar to $Ni(OH)_2$ [80], with the shake-up satellites centered at 861.6 and 879.6 eV. In this sample, the HRXPS spectrum in the region of O 1*s* core level (Figure S19d) consists of two features that can be ascribed to the Ni-OH bond (530.6 eV) and adsorbed water (531.6 eV) [78,80].

The O 1*s* HRXPS spectra of the other monometallic reference oxide (Figure S19c) and of Ni85Co15_400 (Figure 5g) display three distinct features at 529.5, 531.3, and 532.2 eV. They can be attributed to lattice oxygen (O_L), surface oxygen vacancies (O_V), where surface oxygen anions adjacent to lattice vacancies are passivated with hydrogen [41,74,81], and adsorbed O species (O_A), respectively [41,74]. The concentration of surface O_Vs, which can be beneficial to the OER as they act as OH^- adsorption sites [82,83], is slightly higher in Ni85Co15_400 (39.6%) and Ni85Fe15_800 (38.8%, Figure 5i) than in Ni100 (35.0%). Finally, a weak additional contribution, ascribable to adsorbed or chemisorbed O-species such as O_2 or H_2O [41,74], is detected at higher BEs in the O 1*s* HRXPS spectrum of Ni85Fe15_400 (Figure 5h). The very strong intensity of the peak at 531.5 eV reveals that the surface of this sample is highly defective, with 62% O_Vs.

3.2.5. Species on the Surface of HER Electrocatalysts

Figure S20 displays the XPS survey spectra of the reduced oxides evaluated as HER electrocatalysts. Nickel, oxygen, and adventitious carbon are still detected in all electrocatalysts, but compared to the adventitious carbon peak, the intensity of the oxygen peak becomes weaker (compare Figures S18 and S20), which confirms the successful reduction of the oxide surface.

Figure 6 displays HRXPS spectra of Co 2*p*, Ni 2*p*, and O 1*s* core levels in Ni85Co15_400R reduced oxide. The intensity of satellite peaks in the spectrum of Co 2*p* core level (Figure 6a) weakens, and a new peak at ca. 777.8 eV appears alongside those at ca. 786 and 803 eV, proving the formation of metallic Co species on the sample surface [84]. Analogously, in the spectrum of Ni 2*p* core level (Figure 6b), the detection of an additional contribution at ca. 852.2 eV confirms the formation of metallic Ni species [85], in full agreement with the indications provided by Rietveld refinements to the XRD data. Moreover, in the O 1*s* HRXPS spectrum (Figure 6c), the intensity of O_L peak dramatically reduces (25.5%), while that of O_V remarkably intensifies (61.0%), so as their intensity ratio changes from 1.45 to 0.42, confirming the occurrence of partial oxide reduction.

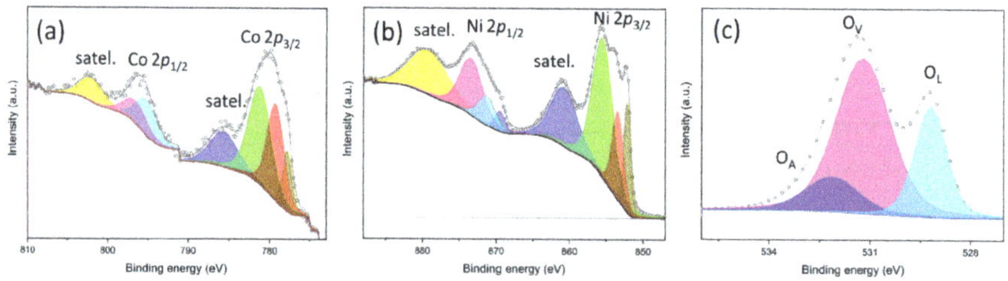

Figure 6. High-resolution XPS curves and fitting in the regions of (**a**) Co 2*p*, (**b**) Ni 2*p*, and (**c**) O 1*s* core levels of Ni85Co15_400R reduced oxide.

After reduction, the contribution of metallic Ni species is detected in all remaining Ni-containing samples examined (Ni100R, NiH100R, Ni85Fe15_400R, and Ni85Fe15_800R), even if with different intensities. In the O 1*s* spectra of samples Ni100R (Figure S21f) and Ni85Fe15_800R (Figure S21t), a significant weakening of the lowest BE peaks signals the diminishing of O_L-species concentration, again confirming the partial reduction of the sample surface. No similar change is observed in sample Ni85Fe15_400R (Figure S21r), whose surface is probably reduced to a little extent, in agreement with the indications that emerged from Rietveld refinements to XRD data (Table 3). No obvious change is observed in the Fe 2*p* spectra (Figure S21m–p), indicating that iron is hardly reduced. Finally, in the O 1*s* spectrum of sample NiH100R (Figure S21h), the peak undergoes a decrease in intensity and a shift towards lower BEs.

3.3. Electrochemical Performance of the Catalysts

As mentioned above, electrochemical tests were performed on five selected pairs of samples. Their choice was based on the search for the best compromise between high phase purity and small average crystal size. Table 4 reports the codes of MEAs fabricated using them.

Table 4. Codes and composition of the membrane-electrode assemblies (MEAs) electrochemically tested. Percentage of ionomer: 20%, for both electrodes. Ionomer: Fumatech® ION FAA3 (10%). Membrane: Fumatech®FAA3-50. Value of cell potential (V_C) and current density (J) achieved by all MEAs at fixed J (0.9 A cm^{-2}) and V_C (2.2 V), respectively.

MEA Code and Composition		Cathode	Anode	V_C/V	J/A cm^{-2}
1	NiH100-based	NiH100R	NiH100	2.13	1.08
2	Ni100-based	Ni100R	Ni100	2.14	1.09
3	Ni85Co15_400-based	Ni85Co15_400R	Ni85Co15_400	2.03	1.78
4	Ni85Fe15_400-based	Ni85Fe15_400R	Ni85Fe15_400	2.20	0.90
5	Ni85Fe15_800-based	Ni85Fe15_800R	Ni85Fe15_800	2.20	0.90

After assembling the MEAs in a single cell, a short (two hours) conditioning was carried out in galvanostatic mode at low current density (50 mA cm^{-2}) and 50 °C in order to promote their activation (Figure S22). After initial variation, the cell potential remained quite stable for all MEAs. Only a slight increase over time was observed for the MEAs (MEA 1 and MEA 2) based on reference monometallic oxides and for (Ni85Fe15_800-based) MEA 5; on the contrary, the cell potential of (Ni85Co15_400-based) MEA 3 and (Ni85Fe15_400-based) MEA 4 slightly decreased. At the end of the conditioning, (Ni100-based) MEA 2 and (Ni85Fe15_800-based) MEA 5 achieved the highest potential (1.84 V), whereas (Ni85Co15_400-based) MEA 3 and (Ni85Fe15_400-based) MEA 4 exhibited lower voltage (1.78 V). The best performance was observed for the MEA based on materials produced by the CP method (NiH100-based) MEA 1 that reached the lowest cell potential (1.74 V) operating at the same current density.

Figure 7a,b compares the polarization curves of the cells investigated at the beginning (BoT, Figure 7a) and end of tests (EoT, Figure 7b). The performance of the present MEAs compares well with those reported in the literature for AEMWE cells (Table S4). At the EoT, all MEAs show an improvement (i.e., higher current density at lower potential) with respect to the BoT. The highest current density at the cut-off voltage (2.2 V) is always achieved by MEA 3: 1 A cm^{-2} at the BoT and 1.78 A cm^{-2} at the EoT. At 50 °C, there are no significant differences in activation losses. The MEA performance at the EoT is ranked (Table 4) by comparing the values of cell potential at a given current density (0.9 A cm^{-2}, i.e., the maximum value reached by MEA 4 and MEA 5 at the cut-off voltage) and those of current density at a given cell potential (2.2 V). The lower the cell voltage for fixed current density (Figure 7c) and the higher the current for fixed cell voltage (Figure 7d), the better the cell electrochemical performance. Based on these criteria, the performance improves in the order MEA 4 = MEA 5 < MEA 2 < MEA 1 < MEA 3.

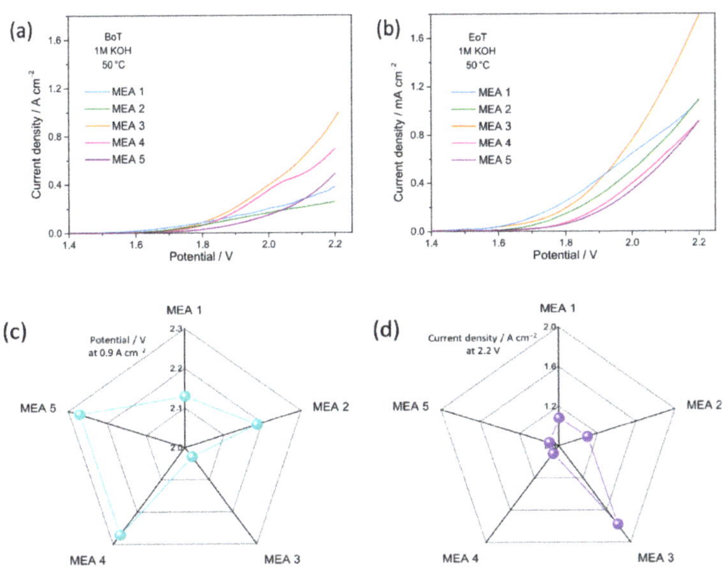

Figure 7. (**a**,**b**) AEMWE cell polarization curves. The shown data refer to the (**a**) beginning of test (BoT) and (**b**) end of test (EoT); for an easier comparison, the same scales are used for both plots. (**c**) Cell potential at 0.9 A cm^{-2} and (**d**) current density at 2.2 V.

Although further investigations would be necessary, the above results confirm that, regardless of the synthesis method, the phase purity (Figure S17a,b) of the anode material has a great influence on the MEA performance. Higher current densities and/or lower potentials are generally obtained with highly pure OER electrocatalysts, in agreement with recent literature reports [71]. Instead, the phase purity of the cathode material seems to be less critical than that of the anodic one, in line with the widely agreed assessment that OER represents the bottleneck of the WE process. In fact, among MEAs with a single-phase anode material (MEA 1-3), the highest current density (~1.8 A cm^{-2}) pertains to MEA 3, whose cathode material has only 93% purity rather than to those with 100% purity (Figure S17).

Also, the synergy between metals seems to contribute to enhancing the electrochemical performance, both in terms of HER and OER [86]. Actually, in the present case, MEA 3 outperforms the two MEAs based on the monometallic reference electrocatalysts. However, the impact of the synergy appears to be lower than that of phase-purity, since MEA 4 and MEA 5, pairing low-purity anode (77–81%) and cathode (72–91%) materials, are able to deliver relatively low current density (<1 A cm^{-2}) despite their bimetallic nature.

The crucial role played by reactive surface defects in boosting the OER has been frequently highlighted in the literature. Recent studies have pointed out that both surface hydroxide species [74] and O-vacancies play a beneficial role towards OER [41], while smaller-sized crystals generally allow shorter charge migration paths on their surface [41]. Indeed, large amounts of $Co(OH)_2$ and $Ni(OH)_2$ are detected by XPS on the highly defective surface of the finely grained Ni85Co15_400, together with a high concentration of O_Vs (39.6%). In Ni85Fe15_800, on whose surface a comparable O_V-amount (38.8%) is present, oxide crystals are larger (55 against 42 nm), and this may hamper the charge migration. As O_Vs behave as a descriptor for the OER process [41], an optimal range of concentration exists. This explains why the huge Ov-amount (62%) in Ni85Fe15_400 ultimately proves detrimental, and the electrochemical performance of MEA 4 is unsatisfactory, despite the relatively small crystal size (28 nm) could be beneficial for charge migration.

Figure 8a–d compares EIS spectra recorded on all MEAs at 50 °C at 1.8 and 2.0 V cell potential. At higher potential (2 V), clear evidence of at least two distinct semicircles appears in the Nyquist plots. The semicircle with low frequency, attributed to the cathode, is significantly larger than one with high frequency, which is associated with the anode. After the durability test (Figure 8d), the semicircles appear significantly overlapped. It is possible that the anode semicircle shifts towards lower frequencies due to reversible losses at the anode.

In order to more deeply understand the electrochemical performance described above, the series resistance (R_S) and polarization resistance (R_P) were estimated from the Nyquist plots, as reported in Section 2.5. The MEA activation during operation, a process involving both catalysts and membrane, can be responsible for the observed enhancement in cell performance. Improved interfacial contact between the membrane and the electrode likely contributes as well. The significant decrease in the values of both R_S and R_P supports these hypotheses for most of the MEAs.

Regardless of the cell potential (1.8 or 2.0 V), MEA 3 shows lower R_S than other MEAs (Figure 8e,g), both before and after the tests. Therefore, MEA 3 is the most efficient. This behavior confirms that the MEA performance is greatly influenced by the phase purity and average crystal size of the electrode materials. Although further investigations would be necessary, the synthesis method seems to affect the performance: Under the same conditions, all MEAs prepared using materials synthesized via the SG method show lower R_S than MEA based on materials produced by the CP method (NiH100-based MEA 1).

The R_P values displayed in Figure 8f,h reveal that MEA 3 generally exhibits a better interface with the membrane. Forming a proper interface with the membrane decreases the mass transfer constraints due to the accumulation of developed O_2 in the catalytic layer.

Figure 9 displays the variation undergone by the cell voltage during the durability tests, carried out at 50 °C by recirculating a 1 M KOH solution to the anode side. The values of current densities, cell voltages, and durability times can be found in Table S5. Only MEA 3 is able to reach a current density of 1 A cm^{-2}. The remaining ones operate at lower current densities, with their performance progressively degrading over time until reaching the set cut-off voltage (2.2 V). Reversible losses are observed during start-up and stop cycles. A possible explanation could be the occurrence of some diffusion limitations that influence the release of the product gas. Due to reversible losses during cycles, all durability test curves present a slightly lower voltage efficiency than steady-state operation.

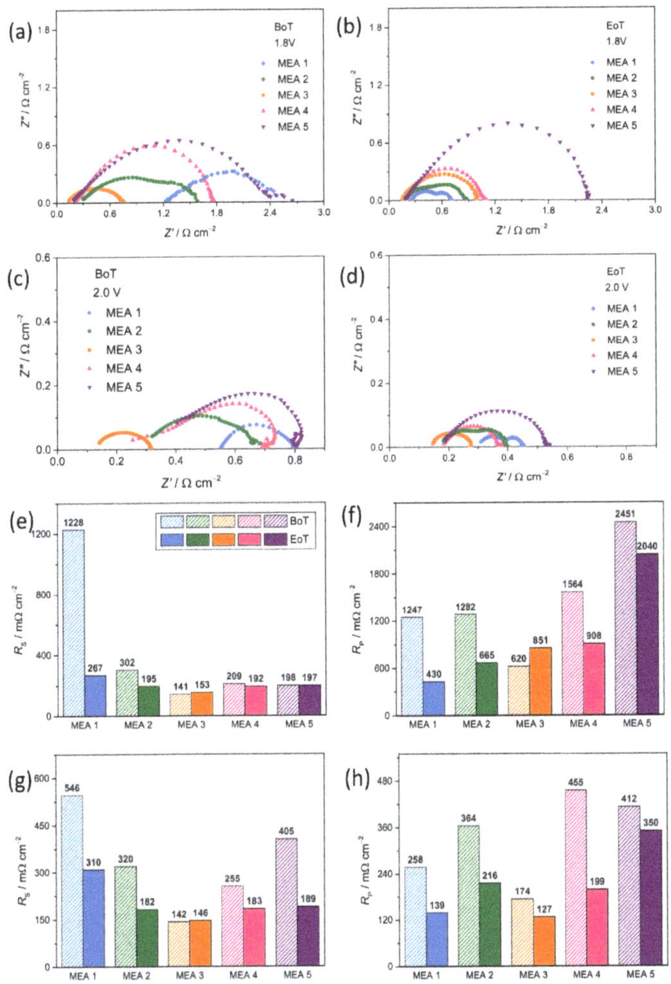

Figure 8. (**a–d**) Nyquist plots at different potentials: (**a**,**b**) 1.8 V and (**c**,**d**) 2.0 V. The shown data refer to the (**a**,**c**) beginning of test (BoT) and (**b**,**d**) end of test (EoT). At each potential, the same scales are used for an easier comparison. (**e**,**g**) R_S and (**f**,**h**) R_P values at (**e**,**f**) 1.8 V and (**g**,**h**) 2.0 V at BoT and EoT.

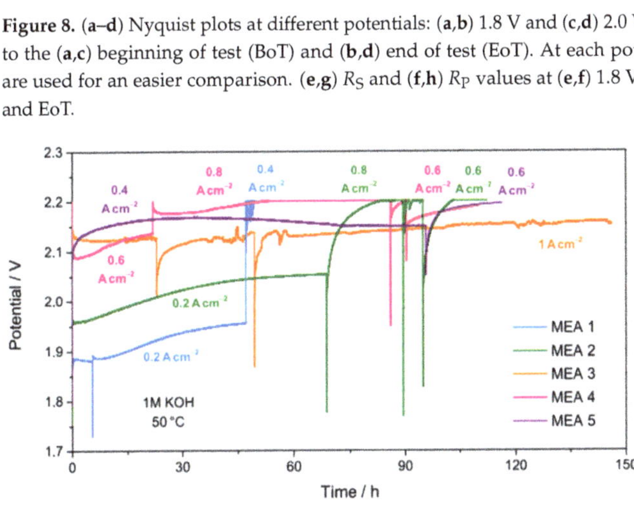

Figure 9. Durability tests at 50 °C and ambient pressure for all MEAs.

By comparing the values of current densities, cell voltages and durability times reported in Table S5, it emerges that the only MEA prepared using materials produced by the CP method (MEA 1) reaches the cut-off voltage (2.2 V) at low current density (0.4 A cm^{-2}) after solely 48 h of durability test. For MEA 2, MEA 4, and MEA 5, durability time is longer (120 h), but they never reach a current density higher than 0.8 A cm^{-2}. The best performance (1 A cm^{-2} at 2.15 V), during a steady-state 150 h-long stability test with 1 M KOH recirculating through the cell, is achieved with the MEA 3 that outperforms the remaining ones also in this test. Hence, Ni85Co15_400-based MEA is the most successful in a zero-gap AEMWE full cell.

3.4. Structural Stability of the Electrocatalysts

The structural stability of all electrocatalysts was evaluated by carrying out XRD analysis on the fresh and reacted electrodes (Figure S23). The signals arising from the electrode supports are clearly observed both before and after the reaction. In addition to the XRD peaks from the Ni-felt, the reflections from the crystallographic planes of the reference β-$Ni(OH)_2$ and NiO electrocatalysts are detected in both the fresh and reacted anodes of MEA 1 (Figure S23a) and MEA 2 (Figure S23c), respectively. Analogously, the reflections from the crystallographic planes of the RS-structured single/main component of the catalysts in their oxidized form, clearly visible in the fresh anodes of MEA 3 (Figure S23e) and MEA 4 (Figure S23g), are still detected after reaction. These findings prove the structural integrity of the anode materials: Although, obviously, their surface is modified during the reaction, the bulk structure of the pristine material remains unchanged [87]. Similar considerations apply to cathode materials. In addition to the signal from the carbon paper [88], the diffraction peaks from the CM (single/main component of the catalysts in their reduced form) are detected in both the fresh and reacted cathodes of all MEAs (Figure S23b,d,f,h).

4. Conclusions

A very simple scalable method for the synthesis of efficient and low-cost electrocatalysts for alkaline water electrolysis is proposed. Nanostructured NiCo- and NiFe-oxides are prepared by the sol–gel method and subsequent calcination in air at different temperatures and then evaluated as anode materials in a zero-gap AEMWE full cell. The cell cathodes are fabricated using the same materials after reduction in a H_2/Ar atmosphere.

The results of electrochemical tests clearly point to a significant role played by the nanomaterial phase purity and average crystal size in determining cell performance. Higher current densities at lower overpotentials are obtained using highly pure and finely grained electrocatalysts. The comparison with reference MEAs based on monometallic oxides suggests that the synergistic action between metals can contribute to enhancing the electrochemical performance.

Among those investigated, MEA joining Ni85Co15_400 anode (consisting of ~36 nm-sized 100% purity rock-salt oxide crystals with abundant O_Vs and hydroxide species on their highly defective surface) and Ni85Co15_400R cathode (consisting of ~56 nm-sized 93% purity cubic metal/alloy) outperforms the remaining ones. It exhibits a current density of 1 A cm^{-2} at 2.15 V during a steady-state 150 h long stability test with 1 M KOH recirculating through the cell, the lowest R_S at any cell potential (1.8 or 2.0 V), and the highest current density at the cut-off voltage (2.2 V) both at the BoT (1 A cm^{-2}) and EoT (1.78 A cm^{-2}).

Although there is still room for improvement, the presented results demonstrate, as proof of concept, the possibility of producing green hydrogen via AEMWE using a very simple and scalable process for the synthesis of nanostructured NiCo- and NiFe-based cell electrode materials.

Supplementary Materials: The following supporting information can be downloaded at https://www.mdpi.com/article/10.3390/nano15131042/s1, Figure S1: Schematic description of the experimental procedure followed to prepare nanostructured oxides to be used as OER electrocatalysts at the cell anode; Figure S2: Synthesis scheme of anode electrocatalyst via co-precipitation method; Figure S3: Schematic description of the experimental procedure for the preparation of nanostructured HER electrocatalysts (cathode materials); Table S1: Codes and reduction temperatures (T_R) of each oxide; Figure S4: TPR profiles of the bimetallic oxides in H_2/Ar atmosphere for 30 min; Figure S5: Airbrush system for manual electrodeposition of catalytic inks; Figure S6: Cathode, anode, Fumatech® membrane, and MEA with 5 cm^2 active area; Figure S7: Scheme of a single cell unit used for the electrochemical assessment; Figure S8: SEM micrographs and SEM/EDX spectra of OER electrocatalysts; Figure S9: XRD patterns of reference electrocatalysts; Figure S10: Rietveld refinements from XRD data of OER catalysts; Table S2: Results of Rietveld refinements for OER electrocatalysts; Figure S11: Effect of the change in synthesis parameter in terms of phases formed in bimetallic oxides; Figure S12: Micro-Raman spectra, as measured at different random locations, within monometallic Ni100 reference electrocatalyst; Figure S13: Micro-Raman spectra, as measured at different random locations, within each bimetallic oxide; Figure S14: SEM micrographs of HER electrocatalysts; Figure S15: SEM/EDX spectra of HER electrocatalysts; Figure S16: Rietveld refinements for reduced oxides; Table S3: Results of Rietveld refinements for HER electrocatalyst; Figure S17: Phase purity of electrode materials in the (a) oxidized and (b) reduced form and (c) corresponding average crystal size, as resulting from Rietveld refinements to XRD data; Figure S18: Survey XPS spectra of the oxides evaluated as OER electrocatalysts; Figure S19: High-resolution XPS curves and fitting in the regions of Ni 2*p* and O 1*s* core levels of Ni100 and NiH100 samples; Figure S20: Survey XPS spectra of the reduced oxides evaluated as HER electrocatalysts; Figure S21: Comparison between HRXPS spectra of electrocatalysts in the oxidized and reduced form; Figure S22: Conditioning at low current density (50 mA cm^{-2}) and 50 °C; Table S4: Performance and durability comparison of different membrane–electrode assemblies in similar AEMWE conditions; Table S5: Current densities, cell voltages, and durability test times for AEM systems fed with 1 M KOH; Figure S23: XRD patterns of anodes and cathodes utilized to fabricate MEA 1, MEA 2, MEA 3, and MEA 4 fresh and after reaction. (References [89–92] are cited in Supplementary Materials).

Author Contributions: Conceptualization, S.S. and S.C.Z.; methodology, C.T., C.A., M.F. and M.P.; validation and formal analysis C.T., C.A., M.F. and M.P.; data curation C.T., M.F. and S.S.; writing—original draft preparation S.S., M.G.M. and S.C.Z.; writing—review and editing, S.S., M.G.M. and S.C.Z.; funding acquisition, S.S. and S.C.Z. All authors have read and agreed to the published version of the manuscript.

Funding: This research was funded by the Italian Ministry of University and Research (MUR) and the European Union—NextGenerationEU, grant PRIN 2022 PNRR "High-Entropy Materials as cost-effective catalysts in anion exchange membrane water electrolysis for GREen hydrogEN production—HEM4GREEN", P2022BHKRA.

Data Availability Statement: Data are contained within the article or Supplementary Materials.

Acknowledgments: The authors gratefully acknowledge the Italian Ministry of University and Research (MUR) for the financial support to this study through the National Recovery and Resilience Plan (NRRP)—Mission 4 "Education and Research"—C2 component—Investment 1.1, funds for the National Research Program and Projects of Relevant National Interest (NRP), P2022BHKRA, funded by the European Union—NextGenerationEU.

Conflicts of Interest: The authors declare no conflicts of interest.

References

1. IEA. *World Energy Outlook 2024*; IEA: Paris, France, 2024. Available online: https://www.iea.org/reports/world-energy-outlook-2024 (accessed on 11 June 2025).
2. Hassan, Q.; Viktor, P.; Al-Musawi, T.J.; Ali, B.M.; Algburi, S.; Alzoubi, H.M.; Al-Jiboory, A.K.; Sameen, A.Z.; Salman, H.M.; Jaszczur, M. The Renewable Energy Role in the Global Energy Transformations. *Renew. Energy Focus* **2024**, *48*, 100545. [CrossRef]
3. Algburi, S.; Al-Dulaimi, O.; Fakhruldeen, H.F.; Khalaf, D.H.; Hanoon, R.N.; Jabbar, F.I.; Hassan, Q.; Al-Jiboory, A.K.; Kiconco, S. The Green Hydrogen Role in the Global Energy Transformations. *Renew. Sustain. Energy Transit.* **2025**, *8*, 100118. [CrossRef]
4. Le, P.-A.; Trung, V.D.; Nguyen, P.L.; Bac Phung, T.V.; Natsuki, J.; Natsuki, T. The Current Status of Hydrogen Energy: An Overview. *RSC Adv.* **2023**, *13*, 28262–28287. [CrossRef] [PubMed]
5. Agrawal, D.; Mahajan, N.; Singh, S.A.; Sreedhar, I. Green Hydrogen Production Pathways for Sustainable Future with Net Zero Emissions. *Fuel* **2024**, *359*, 130131. [CrossRef]
6. Bhuiyan, M.M.H.; Siddique, Z. Hydrogen as an Alternative Fuel: A Comprehensive Review of Challenges and Opportunities in Production, Storage, and Transportation. *Int. J. Hydrogen Energy* **2025**, *102*, 1026–1044. [CrossRef]
7. Kumar, S.S.; Lim, H. An Overview of Water Electrolysis Technologies for Green Hydrogen Production. *Energy Rep.* **2022**, *8*, 13793–13813. [CrossRef]
8. Mulk, W.U.; Aziz, A.R.A.; Ismael, M.A.; Ghoto, A.A.; Ali, S.A.; Younas, M.; Gallucci, F. Electrochemical Hydrogen Production through Anion Exchange Membrane Water Electrolysis (AEMWE): Recent Progress and Associated Challenges in Hydrogen Production. *Int. J. Hydrogen Energy* **2024**, *94*, 1174–1211. [CrossRef]
9. Vedrtnam, A.; Kalauni, K.; Pahwa, R. A Review of Water Electrolysis Technologies with Insights into Optimization and Numerical Simulations. *Int. J. Hydrogen Energy* **2025**, *140*, 694–727. [CrossRef]
10. Miller, H.A.; Bouzek, K.; Hnat, J.; Loos, S.; Bernäcker, C.I.; Weißgärber, T.; Röntzsch, L.; Meier-Haack, J. Green Hydrogen from Anion Exchange Membrane Water Electrolysis: A Review of Recent Developments in Critical Materials and Operating Conditions. *Sustain. Energy Fuels* **2020**, *4*, 2114–2133. [CrossRef]
11. Santos, D.M.F.; Sequeira, C.A.C.; Figueiredo, J.L. Hydrogen Production by Alkaline Water Electrolysis. *Quím. Nova* **2013**, *36*, 1176–1193. [CrossRef]
12. In Lee, H.; Dung, D.T.; Kim, J.; Pak, J.H.; Kim, S.K.; Cho, H.S.; Cho, W.C.; Kim, C.H. The Synthesis of a Zirfon-type Porous Separator with Reduced Gas Crossover for Alkaline Electrolyzer. *Int. J. Energy Res.* **2020**, *44*, 1875–1885. [CrossRef]
13. Carmo, M.; Fritz, D.L.; Mergel, J.; Stolten, D. A Comprehensive Review on PEM Water Electrolysis. *Int. J. Hydrogen Energy* **2013**, *38*, 4901–4934. [CrossRef]
14. Kumar, S.S.; Himabindu, V. Hydrogen Production by PEM Water Electrolysis—A Review. *Mater. Sci. Energy Technol.* **2019**, *2*, 442–454. [CrossRef]
15. Zhang, K.; Liang, X.; Wang, L.; Sun, K.; Wang, Y.; Xie, Z.; Wu, Q.; Bai, X.; Hamdy, M.S.; Chen, H.; et al. Status and Perspectives of Key Materials for PEM Electrolyzer. *Nano Res. Energy* **2022**, *1*, e9120032. [CrossRef]
16. Park, J.E.; Kang, S.Y.; Oh, S.-H.; Kim, J.K.; Lim, M.S.; Ahn, C.-Y.; Cho, Y.-H.; Sung, Y.-E. High-Performance Anion-Exchange Membrane Water Electrolysis. *Electrochim. Acta* **2019**, *295*, 99–106. [CrossRef]
17. Du, N.; Roy, C.; Peach, R.; Turnbull, M.; Thiele, S.; Bock, C. Anion-Exchange Membrane Water Electrolyzers. *Chem. Rev.* **2022**, *122*, 11830–11895. [CrossRef]
18. Hua, D.; Huang, J.; Fabbri, E.; Rafique, M.; Song, B. Development of Anion Exchange Membrane Water Electrolysis and the Associated Challenges: A Review. *ChemElectroChem* **2023**, *10*, e202200999. [CrossRef]
19. Vincent, I.; Bessarabov, D. Low Cost Hydrogen Production by Anion Exchange Membrane Electrolysis: A Review. *Renew. Sustain. Energy Rev.* **2018**, *81*, 1690–1704. [CrossRef]
20. Lee, S.A.; Kim, J.; Kwon, K.C.; Park, S.H.; Jang, H.W. Anion Exchange Membrane Water Electrolysis for Sustainable Large-scale Hydrogen Production. *Carbon Neutralization* **2022**, *1*, 26–48. [CrossRef]
21. Volk, E.K.; Kreider, M.E.; Kwon, S.; Alia, S.M. Recent Progress in Understanding the Catalyst Layer in Anion Exchange Membrane Electrolyzers—Durability, Utilization, and Integration. *EES Catal.* **2024**, *2*, 109–137. [CrossRef]
22. You, B.; Sun, Y. Innovative Strategies for Electrocatalytic Water Splitting. *Acc. Chem. Res.* **2018**, *51*, 1571–1580. [CrossRef] [PubMed]
23. Li, X.; Hao, X.; Abudula, A.; Guan, G. Nanostructured Catalysts for Electrochemical Water Splitting: Current State and Prospects. *J. Mater. Chem. A* **2016**, *4*, 11973–12000. [CrossRef]
24. Faid, A.Y.; Barnett, A.O.; Seland, F.; Sunde, S. Highly Active Nickel-Based Catalyst for Hydrogen Evolution in Anion Exchange Membrane Electrolysis. *Catalysts* **2018**, *8*, 614. [CrossRef]
25. Darband, G.B.; Aliofkhazraei, M.; Rouhaghdam, A.S. Facile Electrodeposition of Ternary Ni-Fe-Co Alloy Nanostructure as a Binder Free, Cost-Effective and Durable Electrocatalyst for High-Performance Overall Water Splitting. *J. Colloid Interface Sci.* **2019**, *547*, 407–420. [CrossRef]

26. Park, Y.S.; Jeong, J.; Noh, Y.; Jang, M.J.; Lee, J.; Lee, K.H.; Lim, D.C.; Seo, M.H.; Kim, W.B.; Yang, J.; et al. Commercial Anion Exchange Membrane Water Electrolyzer Stack through Non-Precious Metal Electrocatalysts. *Appl. Catal. B Environ.* **2021**, *292*, 120170. [CrossRef]
27. Faid, A.Y.; Barnett, A.O.; Seland, F.; Sunde, S. NiCu Mixed Metal Oxide Catalyst for Alkaline Hydrogen Evolution in Anion Exchange Membrane Water Electrolysis. *Electrochim. Acta* **2021**, *371*, 137837. [CrossRef]
28. Campagna Zignani, S.; Lo Faro, M.; Trocino, S.; Aricò, A.S. Investigation of NiFe-Based Catalysts for Oxygen Evolution in Anion-Exchange Membrane Electrolysis. *Energies* **2020**, *13*, 1720. [CrossRef]
29. Sun, H.; Yan, Z.; Liu, F.; Xu, W.; Cheng, F.; Chen, J. Self-Supported Transition-Metal-Based Electrocatalysts for Hydrogen and Oxygen Evolution. *Adv. Mater.* **2020**, *32*, 1806326. [CrossRef]
30. Campagna Zignani, S.; Faro, M.L.; Carbone, A.; Italiano, C.; Trocino, S.; Monforte, G.; Aricò, A.S. Performance and Stability of a Critical Raw Materials-Free Anion Exchange Membrane Electrolysis Cell. *Electrochim. Acta* **2022**, *413*, 140078. [CrossRef]
31. Bhat, K.S.; Nagaraja, H.S. Recent Trends and Insights in Nickel Chalcogenide Nanostructures for Water-Splitting Reactions. *Mater. Res. Innov.* **2021**, *25*, 29–52. [CrossRef]
32. Xue, Y.; Liu, M.; Qin, Y.; Zhang, Y.; Zhang, X.; Fang, J.; Zhang, X.; Zhu, W.; Zhuang, Z. Ultrathin NiFeS Nanosheets as Highly Active Electrocatalysts for Oxygen Evolution Reaction. *Chin. Chem. Lett.* **2022**, *33*, 3916–3920. [CrossRef]
33. Kandel, M.R.; Pan, U.N.; Paudel, D.R.; Dhakal, P.P.; Kim, N.H.; Lee, J.H. Hybridized Bimetallic Phosphides of Ni–Mo, Co–Mo, and Co–Ni in a Single Ultrathin-3D-Nanosheets for Efficient HER and OER in Alkaline Media. *Compos. Part B Eng.* **2022**, *239*, 109992. [CrossRef]
34. Park, D.-H.; Kim, M.-H.; Kim, M.; Byeon, J.-H.; Jang, J.-S.; Kim, J.-H.; Lim, D.-M.; Park, S.-H.; Gu, Y.-H.; Kim, J.; et al. Spherical Nickel Doped Cobalt Phosphide as an Anode Catalyst for Oxygen Evolution Reaction in Alkaline Media: From Catalysis to System. *Appl. Catal. B Environ.* **2023**, *327*, 122444. [CrossRef]
35. Batool, M.; Hameed, A.; Nadeem, M.A. Recent Developments on Iron and Nickel-Based Transition Metal Nitrides for Overall Water Splitting: A Critical Review. *Coord. Chem. Rev.* **2023**, *480*, 215029. [CrossRef]
36. Moschkowitsch, W.; Zion, N.; Honig, H.C.; Levy, N.; Cullen, D.A.; Elbaz, L. Mixed-Metal Nickel–Iron Oxide Aerogels for Oxygen Evolution Reaction. *ACS Catal.* **2022**, *12*, 12162–12169. [CrossRef]
37. Van Limpt, R.T.M.; Lao, M.; Tsampas, M.N.; Creatore, M. Unraveling the Role of the Stoichiometry of Atomic Layer Deposited Nickel Cobalt Oxides on the Oxygen Evolution Reaction. *Adv. Sci.* **2024**, *11*, 2405188. [CrossRef]
38. Patriarchi, A.; Triolo, C.; Minnetti, L.; Muñoz-Márquez, M.Á.; Nobili, F.; Santangelo, S. Evaluation of High-Entropy (Cr, Mn, Fe, Co, Ni)-Oxide Nanofibers and Nanoparticles as Passive Fillers for Solid Composite Electrolytes. *Electrochim. Acta* **2025**, *512*, 145425. [CrossRef]
39. Elhousseini, M.H.; Isık, T.; Kap, Ö.; Verpoort, F.; Horzum, N. Dual Remediation of Waste Waters from Methylene Blue and Chromium (VI) Using Thermally Induced ZnO Nanofibers. *Appl. Surf. Sci.* **2020**, *514*, 145939. [CrossRef]
40. Pantò, F.; Dahrouch, Z.; Saha, A.; Patanè, S.; Santangelo, S.; Triolo, C. Photocatalytic Degradation of Methylene Blue Dye by Porous Zinc Oxide Nanofibers Prepared via Electrospinning: When Defects Become Merits. *Appl. Surf. Sci.* **2021**, *557*, 149830. [CrossRef]
41. Vezzù, K.; Triolo, C.; Moulaee, K.; Pagot, G.; Ponti, A.; Pinna, N.; Neri, G.; Santangelo, S.; Di Noto, V. Interplay Between Calcination Temperature and Alkaline Oxygen Evolution of Electrospun High-Entropy $(Cr_{1/5}Mn_{1/5}Fe_{1/5}Co_{1/5}Ni_{1/5})_3O_4$ Nanofibers. *Small* **2025**, *21*, 2408319. [CrossRef]
42. Zou, D.; Yi, Y.; Song, Y.; Guan, D.; Xu, M.; Ran, R.; Wang, W.; Zhou, W.; Shao, Z. The $BaCe_{0.16}Y_{0.04}Fe_{0.8}O_{3-\delta}$ Nanocomposite: A New High-Performance Cobalt-Free Triple-Conducting Cathode for Protonic Ceramic Fuel Cells Operating at Reduced Temperatures. *J. Mater. Chem. A* **2022**, *10*, 5381–5390. [CrossRef]
43. Popa, N.C. The (Hkl) Dependence of Diffraction-Line Broadening Caused by Strain and Size for All Laue Groups in Rietveld Refinement. *J. Appl. Crystallogr.* **1998**, *31*, 176–180. [CrossRef]
44. Casas-Cabanas, M.; Palacín, M.R.; Rodríguez-Carvajal, J. Microstructural Analysis of Nickel Hydroxide: Anisotropic Size versus Stacking Faults. *Powder Diffr.* **2005**, *20*, 334–344. [CrossRef]
45. Li, T.; Dang, N.; Zhang, W.; Liang, W.; Yang, F. Determining the Degree of [001] Preferred Growth of $Ni(OH)_2$ Nanoplates. *Nanomaterials* **2018**, *8*, 991. [CrossRef]
46. Jayashree, R.S.; Kamath, P.V.; Subbanna, G.N. The Effect of Crystallinity on the Reversible Discharge Capacity of Nickel Hydroxide. *J. Electrochem. Soc.* **2000**, *147*, 2029. [CrossRef]
47. Tientong, J.; Garcia, S.; Thurber, C.R.; Golden, T.D. Synthesis of Nickel and Nickel Hydroxide Nanopowders by Simplified Chemical Reduction. *J. Nanotechnol.* **2014**, *2014*, 193162. [CrossRef]
48. Dharmaraj, N.; Prabu, P.; Nagarajan, S.; Kim, C.H.; Park, J.H.; Kim, H.Y. Synthesis of Nickel Oxide Nanoparticles Using Nickel Acetate and Poly (Vinyl Acetate) Precursor. *Mater. Sci. Eng. B* **2006**, *128*, 111–114. [CrossRef]
49. Ahmad, T.; Ramanujachary, K.V.; Lofland, S.E.; Ganguli, A.K. Magnetic and Electrochemical Properties of Nickel Oxide Nanoparticles Obtained by the Reverse-Micellar Route. *Solid State Sci.* **2006**, *8*, 425–430. [CrossRef]

50. Salavati-Niasari, M.; Davar, F.; Fereshteh, Z. Synthesis of Nickel and Nickel Oxide Nanoparticles via Heat-Treatment of Simple Octanoate Precursor. *J. Alloys Compd.* **2010**, *494*, 410–414. [CrossRef]
51. Shannon, R.D. Revised Effective Ionic Radii and Systematic Studies of Interatomic Distances in Halides and Chalcogenides. *Acta Cryst. A* **1976**, *32*, 751–767. [CrossRef]
52. Cuong, N.D.; Tran, T.D.; Nguyen, Q.T.; Van Minh Hai, H.; Hoa, T.T.; Quang, D.T.; Klysubun, W.; Tran, P.D. Highly Porous Co-Doped NiO Nanorods: Facile Hydrothermal Synthesis and Electrocatalytic Oxygen Evolution Properties. *R. Soc. Open sci.* **2021**, *8*, 202352. [CrossRef] [PubMed]
53. Tharayil, N.J.; Sagar, S.; Raveendran, R.; Vaidyan, A.V. Dielectric Studies of Nanocrystalline Nickel–Cobalt Oxide. *Phys. B Condens. Matter* **2007**, *399*, 1–8. [CrossRef]
54. Iacomi, F.; Calin, G.; Scarlat, C.; Irimia, M.; Doroftei, C.; Dobromir, M.; Rusu, G.G.; Iftimie, N.; Sandu, A.V. Functional Properties of Nickel Cobalt Oxide Thin Films. *Thin Solid Film.* **2011**, *520*, 651–655. [CrossRef]
55. Yuan, X.; Hu, X.-X.; Ding, X.-L.; Kong, H.-C.; Sha, H.-D.; Lin, H.; Wen, W.; Shen, G.; Guo, Z.; Ma, Z.-F.; et al. Effects of Cobalt Precursor on Pyrolyzed Carbon-Supported Cobalt-Polypyrrole as Electrocatalyst toward Oxygen Reduction Reaction. *Nanoscale Res. Lett.* **2013**, *8*, 478. [CrossRef]
56. Liu, G.; Gao, X.; Wang, K.; He, D.; Li, J. Mesoporous Nickel–Iron Binary Oxide Nanorods for Efficient Electrocatalytic Water Oxidation. *Nano Res.* **2017**, *10*, 2096–2105. [CrossRef]
57. Ravikumar, P.; Kisan, B.; Perumal, A. Enhanced Room Temperature Ferromagnetism in Antiferromagnetic NiO Nanoparticles. *AIP Adv.* **2015**, *5*, 087116. [CrossRef]
58. Liu, D.; Li, D.; Yang, D. Size-Dependent Magnetic Properties of Branchlike Nickel Oxide Nanocrystals. *AIP Adv.* **2017**, *7*, 015028. [CrossRef]
59. Usharani, N.J.; Shringi, R.; Sanghavi, H.; Subramanian, S.; Bhattacharya, S.S. Role of Size, Alio-/Multi-Valency and Non-Stoichiometry in the Synthesis of Phase-Pure High Entropy Oxide (Co, Cu, Mg, Na, Ni, Zn)O. *Dalton Trans.* **2020**, *49*, 7123–7132. [CrossRef]
60. D'Ippolito, V.; Andreozzi, G.B.; Bersani, D.; Lottici, P.P. Raman Fingerprint of Chromate, Aluminate and Ferrite Spinels. *J. Raman Spectrosc.* **2015**, *46*, 1255–1264. [CrossRef]
61. Pantò, F.; Fan, Y.; Frontera, P.; Stelitano, S.; Fazio, E.; Patanè, S.; Marelli, M.; Antonucci, P.; Neri, F.; Pinna, N.; et al. Are Electrospun Carbon/Metal Oxide Composite Fibers Relevant Electrode Materials for Li-Ion Batteries? *J. Electrochem. Soc.* **2016**, *163*, A2930–A2937. [CrossRef]
62. Ercolino, G.; Grodzka, A.; Grzybek, G.; Stelmachowski, P.; Specchia, S.; Kotarba, A. The Effect of the Preparation Method of Pd-Doped Cobalt Spinel on the Catalytic Activity in Methane Oxidation Under Lean Fuel Conditions. *Top. Catal.* **2017**, *60*, 333–341. [CrossRef]
63. Farhadi, S.; Javanmard, M.; Nadri, G. Characterization of Cobalt Oxide Nanoparticles Prepared by the Thermal Decomposition of [Co(NH3)5(H2O)](NO3)3 Complex and Study of Their Photocatalytic Activity. *Acta Chim. Slov.* **2016**, *63*, 335–343. [CrossRef]
64. Santangelo, S.; Fiore, M.; Pantò, F.; Stelitano, S.; Marelli, M.; Frontera, P.; Antonucci, P.; Longoni, G.; Ruffo, R. Electro-Spun Co_3O_4 Anode Material for Na-Ion Rechargeable Batteries. *Solid State Ion.* **2017**, *309*, 41–47. [CrossRef]
65. Venkatachalam, V.; Alsalme, A.; Alswieleh, A.; Jayavel, R. Shape Controlled Synthesis of Rod-like Co_3O_4 Nanostructures as High-Performance Electrodes for Supercapacitor Applications. *J. Mater. Sci. Mater. Electron.* **2018**, *29*, 6059–6067. [CrossRef]
66. Ahlawat, A.; Sathe, V.G. Raman Study of $NiFe_2O_4$ Nanoparticles, Bulk and Films: Effect of Laser Power. *J. Raman Spectrosc.* **2011**, *42*, 1087–1094. [CrossRef]
67. Silva, M.M.S.; Raimundo, R.A.; Ferreira, L.S.; Macedo, D.A.; Morales, M.A.; Souza, C.P.; Santos, A.G.; Lopes-Moriyama, A.L. Effects of Morphology on the Electrochemical Performance of $NiFe_2$ O_4 Nanoparticles with Battery-type Behavior. *Int. J. Appl. Ceram. Tech.* **2022**, *19*, 2016–2028. [CrossRef]
68. Ali, B.; Tasirin, S.; Aminayi, P.; Yaakob, Z.; Ali, N.; Noori, W. Non-Supported Nickel-Based Coral Sponge-Like Porous Magnetic Alloys for Catalytic Production of Syngas and Carbon Bio-Nanofilaments via a Biogas Decomposition Approach. *Nanomaterials* **2018**, *8*, 1053. [CrossRef]
69. Jacob, G.A.; Sellaiyan, S.; Uedono, A.; Joseyphus, R.J. Magnetic Properties of Metastable Bcc Phase in Fe64Ni36 Alloy Synthesized through Polyol Process. *Appl. Phys. A* **2020**, *126*, 120. [CrossRef]
70. Senthamizhan, A.; Balusamy, B.; Aytac, Z.; Uyar, T. Grain Boundary Engineering in Electrospun ZnO Nanostructures as Promising Photocatalysts. *CrystEngComm* **2016**, *18*, 6341–6351. [CrossRef]
71. Montalto, M.; Freitas, W.D.S.; Mastronardo, E.; Ficca, V.C.A.; Placidi, E.; Baglio, V.; Mosca, E.; Vecchio, C.L.; Gatto, I.; Mecheri, B.; et al. Spinel-Type High-Entropy Oxides for Enhanced Oxygen Evolution Reaction Activity in Anion Exchange Membrane Water Electrolyzers. *Chem. Eng. J.* **2025**, *507*, 160641. [CrossRef]
72. Zhou, Z.; Zhang, Y.; Wang, Z.; Wei, W.; Tang, W.; Shi, J.; Xiong, R. Electronic Structure Studies of the Spinel CoFe2O4 by X-Ray Photoelectron Spectroscopy. *Appl. Surf. Sci.* **2008**, *254*, 6972–6975. [CrossRef]

73. Chuang, T.J.; Brundle, C.R.; Rice, D.W. Interpretation of the X-Ray Photoemission Spectra of Cobalt Oxides and Cobalt Oxide Surfaces. *Surf. Sci.* **1976**, *59*, 413–429. [CrossRef]
74. Triolo, C.; Moulaee, K.; Bellato, F.; Pagot, G.; Ponti, A.; Maida, S.; Pinna, N.; Neri, G.; Di Noto, V.; Santangelo, S. Interplay between Alkaline Water Oxidation Temperature, Composition and Performance of Electrospun High-Entropy Non-Equimolar (Cr, Mn, Fe, Co, Ni) Oxide Electrocatalysts. *J. Power Sources* 2025, *submitted*.
75. Zhang, W.; Jiang, C.; Guan, H.; Wang, Y.; Hu, Y.; Wang, W.; Tian, W.; Hao, L. Unlocking OER Catalytic Potential and Chiral $Fe_3 O_4$ Film as a Game-Changer for Electrochemical Water Oxidation Pathway and by-Product Control. *Mater. Adv.* **2024**, *5*, 1340–1347. [CrossRef]
76. Sulemana, H.; Yi, C.; Nawaz, M.I.; Zhang, B.; Yi, R.; Zhang, J.; Nkudede, E. Synthesis and Characterization of Nickel Ferrite (NiFe2O4) Nano-Catalyst Films for Ciprofloxacin Degradation. *Ceram. Int.* **2025**, *51*, 8376–8387. [CrossRef]
77. Tomellini, M. X-Ray Photoelectron Spectra of Defective Nickel Oxide. *J. Chem. Soc. Faraday Trans. 1* **1988**, *84*, 3501. [CrossRef]
78. Kim, K.S.; Winograd, N. X-Ray Photoelectron Spectroscopic Studies of Nickel-Oxygen Surfaces Using Oxygen and Argon Ion-Bombardment. *Surf. Sci.* **1974**, *43*, 625–643. [CrossRef]
79. Yang, Y.; Tao, Q.; Srinivasan, G.; Takoudis, C.G. Cyclic Chemical Vapor Deposition of Nickel Ferrite Thin Films Using Organometallic Precursor Combination. *ECS J. Solid State Sci. Technol.* **2014**, *3*, P345–P352. [CrossRef]
80. Yang, Z.; Chen, J.; Liang, Z.; Xie, W.; Zhao, B.; He, L. Anodic Product-Derived Bi-MOF as Pre-catalyst for Cathodic CO_2 Reduction: A Novel Strategy for Paired Electrolysis. *ChemCatChem* **2023**, *15*, e202201321. [CrossRef]
81. Frankcombe, T.J.; Liu, Y. Interpretation of Oxygen 1s X-Ray Photoelectron Spectroscopy of ZnO. *Chem. Mater.* **2023**, *35*, 5468–5474. [CrossRef]
82. Zhu, K.; Shi, F.; Zhu, X.; Yang, W. The Roles of Oxygen Vacancies in Electrocatalytic Oxygen Evolution Reaction. *Nano Energy* **2020**, *73*, 104761. [CrossRef]
83. He, L.; Kang, H.; Hou, G.; Qiao, X.; Jia, X.; Qin, W.; Wu, X. Low-Temperature Synthesis of Nano-Porous High Entropy Spinel Oxides with High Grain Boundary Density for Oxygen Evolution Reaction. *Chem. Eng. J.* **2023**, *460*, 141675. [CrossRef]
84. McIntyre, N.S.; Johnston, D.D.; Coatsworth, L.L.; Davidson, R.D.; Brown, J.R. X-ray Photoelectron Spectroscopic Studies of Thin Film Oxides of Cobalt and Molybdenum. *Surf. Interface Anal.* **1990**, *15*, 265–272. [CrossRef]
85. Furstenau, R.P.; McDougall, G.; Langell, M.A. Initial Stages of Hydrogen Reduction of NiO(100). *Surf. Sci.* **1985**, *150*, 55–79. [CrossRef]
86. Huang, K.; Xia, J.; Lu, Y.; Zhang, B.; Shi, W.; Cao, X.; Zhang, X.; Woods, L.M.; Han, C.; Chen, C.; et al. Self-Reconstructed Spinel Surface Structure Enabling the Long-Term Stable Hydrogen Evolution Reaction/Oxygen Evolution Reaction Efficiency of FeCoNiRu High-Entropy Alloyed Electrocatalyst. *Adv. Sci.* **2023**, *10*, 2300094. [CrossRef]
87. Triolo, C.; Moulaee, K.; Ponti, A.; Pagot, G.; Di Noto, V.; Pinna, N.; Neri, G.; Santangelo, S. Spinel-Structured High-Entropy Oxide Nanofibers as Electrocatalysts for Oxygen Evolution in Alkaline Solution: Effect of Metal Combination and Calcination Temperature. *Adv. Funct. Mater.* **2024**, *34*, 2306375. [CrossRef]
88. Trocino, S.; Lo Vecchio, C.; Campagna Zignani, S.; Carbone, A.; Saccà, A.; Baglio, V.; Gómez, R.; Aricò, A.S. Dry Hydrogen Production in a Tandem Critical Raw Material-Free Water Photoelectrolysis Cell Using a Hydrophobic Gas-Diffusion Backing Layer. *Catalysts* **2020**, *10*, 1319. [CrossRef]
89. Vincent, I.; Kruger, A.; Bessarabov, D. Development of Efficient Membrane Electrode Assembly for Low Cost Hydrogen Production by Anion Exchange Membrane Electrolysis. *Int. J. Hydrogen Energy* **2017**, *42*, 10752–10761. [CrossRef]
90. Liu, Z.; Sajjad, S.D.; Gao, Y.; Yang, H.; Kaczur, J.J.; Masel, R.I. The Effect of Membrane on an Alkaline Water Electrolyzer. *Int. J. Hydrogen Energy* **2017**, *42*, 29661–29665. [CrossRef]
91. Vincent, I.; Lee, E.-C.; Kim, H.-M. Highly Cost-Effective Platinum-Free Anion Exchange Membrane Electrolysis for Large Scale Energy Storage and Hydrogen Production. *RSC Adv.* **2020**, *10*, 37429–37438. [CrossRef]
92. Najibah, M.; Hnát, J.; Plevová, M.; Dayan, A.; Stojadinovic, J.; Žitka, J.; Bae, C.; Bouzek, K.; Henkensmeier, D. PPS-Reinforced Poly(Terphenylene) Anion-Exchange Membranes with Different Quaternary Ammonium Groups for Use in Water Electrolysers. *J. Membr. Sci.* **2025**, *713*, 123335. [CrossRef]

nanomaterials

MDPI

Article

Self-Reconstructed Metal–Organic Framework-Based Hybrid Electrocatalysts for Efficient Oxygen Evolution

Kunting Cai [1], Weibin Chen [1], Yinji Wan [2], Hsingkai Chu [1], Xiao Hai [1,*] and Ruqiang Zou [1,*]

[1] Beijing Key Laboratory for Theory and Technology of Advanced Battery Materials, School of Materials Science and Engineering, Peking University, No. 5 Yiheyuan Road, Haidian District, Beijing 100871, China

[2] State Key Laboratory of Heavy Oil Processing, China University of Petroleum, No. 18 Fuxue Road, Changping District, Beijing 102249, China

* Correspondence: xiaohai@pku.edu.cn (X.H.); rzou@pku.edu.cn (R.Z.)

Abstract: Refining synthesis strategies for metal–organic framework (MOF)-based catalysts to improve their performance and stability in an oxygen evolution reaction (OER) is a big challenge. In this study, a series of nanostructured electrocatalysts were synthesized through a solvothermal method by growing MOFs and metal–triazolates (METs) on nickel foam (NF) substrates (named MET-M/NF, M = Fe, Co, Cu), and these electrocatalysts could be used directly as OER self-supporting electrodes. Among these electrocatalysts, MET-Fe/NF exhibited the best OER performance, requiring only an overpotential of 122 mV at a current density of 10 mA cm^{-2} and showing remarkable stability over 15 h. The experimental results uncovered that MET-Fe/NF underwent an in situ structural reconstruction, resulting in the formation of numerous iron/nickel (oxy)hydroxides with high OER activity. Furthermore, in a two-electrode water-splitting setup, MET-Fe/NF only required 1.463 V to achieve a current density of 10 mA cm^{-2}. Highlighting its potential for practical applications. This work provides insight into the design and development of efficient MOF-based OER catalysts.

Keywords: metal–organic frameworks; structure reconstruction; oxygen evolution reaction; hybrid electrocatalysts

1. Introduction

The rapid consumption of fossil fuels has led to a series of energy and environmental problems, prompting researchers to develop sustainable and clean energy technologies [1–3]. The electrocatalytic oxygen evolution reaction (OER) plays a critical role in renewable energy conversion and storage technologies, such as metal–air batteries, water splitting, and fuel cells [4–6]. However, the OER process involves a complex four-electron transfer pathway, and its slow kinetics significantly limit the efficiency of energy conversion systems [7,8]. Catalysts can efficiently reduce the overpotential and improve the OER process, so the proper design and development of catalysts are important. Currently, iridium dioxide (IrO_2) and ruthenium dioxide (RuO_2) are considered efficient OER catalysts, but their scarcity, high cost, and low stability under alkaline conditions severely hinder their industrial applications [9–11]. Therefore, a key goal for researchers is to design and develop a cost-effective, highly active, and durable non-precious metal OER electrocatalyst [12–15].

In recent years, metal–organic frameworks (MOFs), which are synthesized through the periodic assembly of organic ligands and metal ions/clusters, have garnered widespread application in the field of electrocatalysis due to their easily tunable structures and functionalities [16–24]. Nevertheless, the limited conductivity and chemical stability of many MOFs pose significant challenges to their direct use in electrocatalytic processes [25–27]. Metal–triazolate (MET), $[M(C_2N_3H_2)_2]_n$, is a new family of MOFs that could be prepared by combining 1H-1,2,3-triazole and divalent metal ions (Fe, Cu, Co, and so on) to give six isostructural metal–triazolates (termed MET-M, M = Fe, Cu, Co, and so on). MET frameworks have permanent porosity and display surface areas as high as some of the most porous zeolites, with one member of this family, MET-Fe, exhibiting significant electrical conductivity [28]. Notably, the in situ growth of MOFs on a conductive substrate can be regarded as a facile strategy to construct self-supported electrodes, which can considerably accelerate electron and mass transfer. In addition, 3D metal foams, such as nickel foam (NF), have previously been reported as a well-known class of electrocatalytic support materials for MOFs [29,30]. Significantly, the transformation of many MOF materials into metal (oxy)hydroxides during the alkaline OER process is a widely observed phenomenon [31–34], suggesting that these newly formed metal hydroxides could serve as catalytically active sites for OER, such as iron oxyhydroxides (FeOOH), nickel oxyhydroxides (NiOOH), and nickel iron-layered double hydroxides (NiFe-LDH) [19,35,36]. Furthermore, single-component catalysts are often limited in their ability to improve performance, suggesting that the design of composite catalysts with multiple active elements may be a more effective approach to enhance overall catalytic efficiency due to a catalytic synergistic effect [37–42]. The interaction between different active sites in such catalysts can refine and accelerate reaction processes, thereby enhancing the catalytic activity [43–45].

Considering the above context, in this study, we have successfully synthesized metal–triazolates, (MET-M, $[M(C_2N_3H_2)_2]_n$, M = Fe, Cu, Co), on nickel foam via a solvothermal method, which can directly serve as self-supporting OER electrodes. The best sample MET-Fe/NF material undergoes in situ restructuring during the OER process, and the resulting mixed catalyst with multiple active components (FeOOH, NiOOH, and NiFe-LDH) exhibits excellent OER electrocatalytic activity. The MET-Fe/NF material achieves a current density of 10 mA cm^{-2} with an overpotential of only 122 mV and a small Tafel slope of 34.5 mV dec^{-1}, outperforming commercial RuO_2, and it maintains good stability over 15 h. In two-electrode water splitting applications, the energy conversion device equipped with this catalyst operates at a low voltage of 1.463 V to continuously produce hydrogen and oxygen at a current density of 10 mA cm^{-2}, demonstrating the promising practical application prospects of the catalyst. This work offers an effective strategy for the design of highly efficient MOF-based electrocatalysts.

2. Experimental

2.1. Materials

In this study, all chemicals and reagents were of analytical grade and used without further purification. $FeCl_2$ (purity: 99%) was purchased from Konoscience Reagent Co., Ltd. (Shanghai, China). $CoCl_2$ (purity: 99%) was purchased from Macklin Reagent Co., Ltd. (Seattle, WA, USA). $Cu(NO_3)_2.xH_2O$ (purity: 99%) was purchased from Beijing Tongguang Fine Chemicals Co., Ltd. (Beijing, China). 1H-1,2,3-triazole (purity: 98%) was purchased from Shanghai D&B Co., Ltd. (Shanghai, China). N,N′-dimethylformamide (DMF) (purity: 99.8%) was purchased from Concord Co., Ltd. (Singapore). Commercial NF (purity: 99%) (thickness: 0.5 mm) was provided by Kunshan Guangjiayuan Co., Ltd. (Kunshan, China).

2.2. Preparation of MET-Fe/NF

First, $FeCl_2$ (8.5 mmol) was dissolved in 50 mL of a DMF solution. The solution was then transferred to a Teflon reactor. After that, 1H-1,2,3-triazole (25.5 mmol) was added to the solution and the NF substrate was immersed in the solution. The samples were then obtained after being placed in an oven at 120 °C for 48 h. Finally, after cooling down to RT, the resulting MET-Fe/NF was washed three times with alcohol and dried at 80 °C for 8 h.

2.3. Preparation of MET-Co/NF and MET-Cu/NF

The preparation procedures of MET-Co/NF and MET-Cu/NF were similar to that of MET-Fe/NF, except that the metal sources were $CoCl_2$ and $Cu(NO_3)_2.xH_2O$, respectively.

2.4. Preparation of RuO_2/NF and Pt/C/NF

For comparison, the commercial RuO_2 dispersion was prepared using ultrasonication with 4 mg of RuO_2, 500 µL ethanol, and 100 µL 5 wt% Nafion. After homogenization, 30 μL of the dispersion was drip-coated on NF (1 cm × 1 cm) at a loading of 0.2 mg cm^{-2}. The preparation procedure of Pt/C/NF was similar to that of RuO_2/NF, except that the metal source was commercial Pt/C.

2.5. Characterization

The crystallographic structures of the materials were determined using a Rigaku SmartLab 9 kW diffractometer (Tokyo, Japan) with Cu Kα radiation (λ = 1.5406 Å). Nitrogen sorption isotherms were measured at 77 K on a Quantachrome Autosorb-IQ gas adsorption analyzer (Boynton Beach, FL, USA). From the adsorption isotherms, the specific surface areas were calculated using the Brunauer–Emmett–Teller (BET) method. The pore size distributions were obtained using the quenched solid density functional theory (QSDFT) method. The microstructure and morphology were examined by using a scanning electron microscope (SEM; Hitachi S-4800 microscope, Tokyo, Japan). The surface characterization of elemental electronic states was measured by the Krayos AXIS Ultra DLD electronic energy spectrometer (Shimadzu, Tokyo, Japan) (XPS; Axis Ultra imaging photoelectron spectrometer with the monochromatic Al Kα line). The mid-infrared absorptive spectrum of samples was determined using an ATR-FTIR spectrometer (Bruker, VERTEX 80V, Berlin, Germany).

2.6. Electrochemical Measurements

The electrocatalytic properties of OER were measured using a three-electrode system on a CHI660 workstation (Chenhua, Shanghai, China), in which NF-based electrodes, Hg/HgO electrodes, and platinum electrodes served as the working, reference, and counter electrodes in a 1 M KOH solution, respectively. The potentials were referred to the reversible hydrogen electrode (RHE) according to the following equation [46]: $E_{RHE} = E_{Hg/HgO}$ + 0.098 V + 0.059 × pH. The linear sweep voltammetry (LSV) polarization curves of the electrodes were tested at a scan rate of 5 mV s^{-1} with 95% iR compensation. The Tafel slopes were derived from polarization curves. CV tests were performed at sweep rates ranging from 20 to 100 mV s^{-1} in a non-Faradaic region (0.912~1.012 V vs. RHE), then double-layer capacitance (C_{dl}) was calculated using the CV results. Electrochemical impedance spectroscopy (EIS) was performed in the frequency range from 100 kHz to 0.1 Hz at open circuit potential. Chronopotentiometry measurement was performed at a current density of 10 mA cm^{-2} without iR compensation. Electrochemical water splitting was measured using a two-electrode system.

3. Results and Discussion

The synthesis method for the materials is illustrated in Figure 1, based on adaptations from the literature references [28]. The process involved utilizing Fe^{2+} ions as the metal source, 1H-1,2,3-triazole as the organic ligand, and DMF as the solvent, with nickel foam serving as the growth substrate for the composite material. The procedure involved a one-pot solvothermal reaction in a reaction vessel at 120 °C for 48 h, culminating in the production of MET-Fe/NF sample material. The chosen organic ligand, 1H-1,2,3-triazole, is characterized by a five-membered ring structure containing three nitrogen atoms, which are differentiated into two types, N1 and N2, based on subtle differences in their chemical environment (Figure S1a). During the reaction, all three nitrogen atoms are involved in coordination with the metal atoms, with each metal center engaging in octahedral coordination with six nitrogen atoms, thereby constructing a three-dimensional MOF framework endowed with a porous architecture (Figure S1b). By changing the metal ions to Co^{2+} or Cu^{2+}, two additional comparative samples, MET-Co/NF and MET-Cu/NF, were synthesized. In contrast to the bare nickel foam (Figure S2a), the surfaces of all synthesized samples were uniformly coated with a layer of powdery substance, with the MET-Fe/NF sample exhibiting a black-brown powder (Figure S2b), the MET-Co/NF sample showing a yellow-brown powder (Figure S2c), and the MET-Cu/NF sample displaying a blue powder (Figure S2d). These initial observations confirm the feasibility and universal applicability of synthesizing MOFs/NF composites through a solvothermal one-pot method. The array of MOFs/NF composite materials thus obtained can be directly used as self-supporting electrocatalytic electrodes (with an effective electrode area of 1 cm^2 square), streamlining the subsequent testing of OER performance as well as the analytical characterization of the catalytic materials.

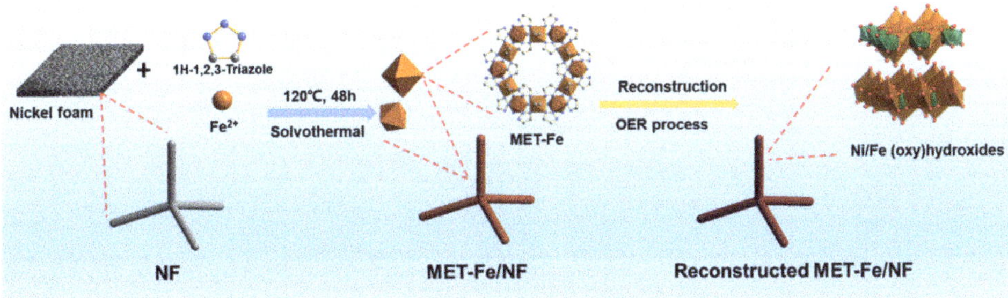

Figure 1. Schematic illustrating the synthesis process and in situ structure reconstruction of the MET-Fe/NF catalyst.

To further determine the phase composition of the samples, X-ray diffraction (XRD) characterization was performed on the samples obtained. As depicted in Figure 2a, the XRD patterns revealed strong peaks corresponding to the nickel foam substrate (PDF# 70-0989), indicating the origin of the Ni peaks. Analysis of the XRD patterns within the 5° to 40° range revealed peaks that were consistent with those in the simulated MET-Fe XRD spectrum, suggesting the successful growth of MET-Fe powder on the surface of the nickel foam. XRD characterization of the precipitate obtained from the one-pot solvothermal reaction, which is homologous to the material on the MET-Fe/NF surface (Figure S3a), further confirmed that the powder on the MET-Fe/NF sample surface is indeed MET-Fe. Phase analysis was also conducted on the comparative samples MET-Co/NF and MET-Cu/NF. Although the XRD peak signals from the nickel foam substrate were dominant (Figure S3b), additional XRD tests on the homologous solids formed during the reaction confirmed that the powders on the surfaces of the comparative samples were MET-Co (Figure S3c) and MET-Cu (Figure S3d), respectively. Furthermore, Brunauer–Emmett–Teller (BET) surface area analysis revealed that the MET-Fe powder on the MET-Fe/NF sample surface possesses a high specific surface area (up to 446 m^2/g) and features a plethora of micropores and mesopores (Figure 2b), which would facilitate extensive contact between the MET-Fe/NF sample and the electrolyte.

Scanning electron microscopy (SEM) was used to investigate the morphology and elemental composition of the samples. Figure S4a shows the nickel foam substrate as a metallic framework with pores on the order of 100 μm. The elemental mapping presented in Figure S4b shows a uniform distribution of oxygen elements over the nickel foam surface. Figure 3a shows that the nickel foam substrate of the MET-Fe/NF sample is thoroughly coated with a solid powder layer, while higher magnification images (Figure 3b) show that this layer is composed of densely packed nano-sized MOF particles. Figure 3c provides a detailed examination of the MET-Fe growth on the nickel foam surface, while the uniform distribution of Ni, O, Fe, and N elements in Figure 3d confirms the uniform coverage of MET-Fe over the nickel foam. SEM characterization was also performed on the MET-Co/NF and MET-Cu/NF samples. Figure S4c shows that there are relatively few solids grown on the surface of MET-Co/NF. In contrast, Figure S5d shows that the bulk solids in MET-Cu/NF fill most of the space within the nickel foam substrate. The different morphological composite structures of the samples may influence the subsequent catalytic processes.

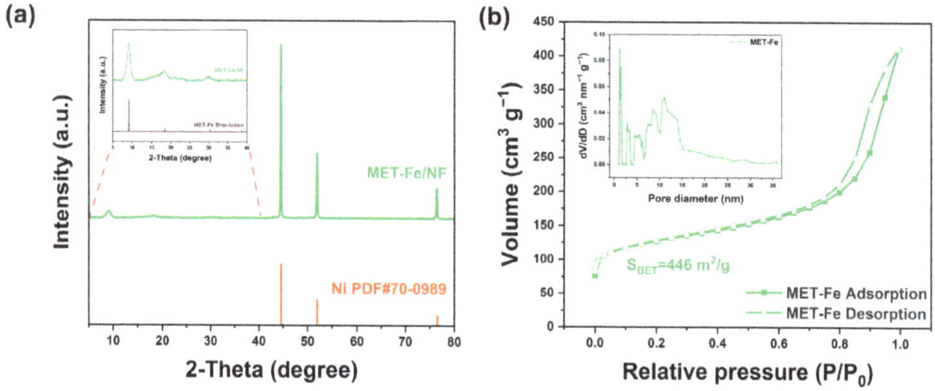

Figure 2. (**a**) XRD pattern of MET-Fe/NF (inset: a detailed view of the dotted area). (**b**) N_2 adsorption-desorption isotherm of MET-Fe (inset: pore size distribution curve).

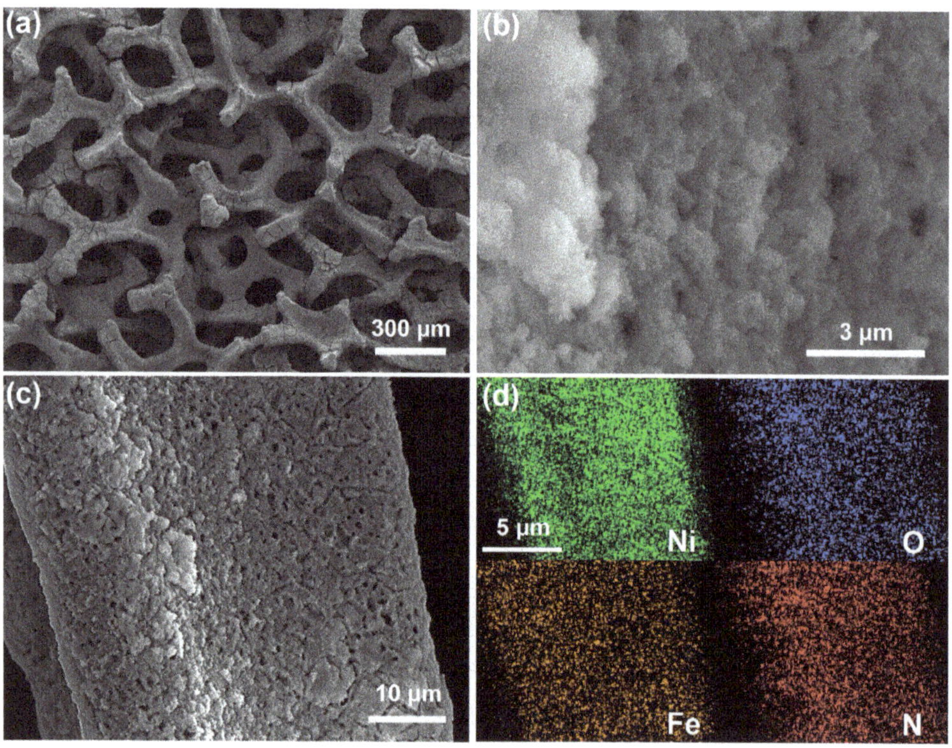

Figure 3. (**a**–**c**) SEM images of MET-Fe/NF at different magnifications. (**d**) Elemental mapping images of MET-Fe/NF.

The OER performance of MET-Fe/NF and reference samples was initially assessed using a standard three-electrode system with 1M KOH as the electrolyte. The linear sweep voltammetry (LSV) curves depicted in Figure 4a indicate that MET-Fe/NF exhibits superior OER performance. As shown in Figure 4b, MET-Fe/NF requires only a 122 mV overpotential to achieve a current density of 10 mA cm^{-2}, outperforming MET-Co/NF (385 mV), MET-Cu/NF (476 mV), bare nickel foam (NF) (375 mV), and the commercial

RuO_2/NF catalyst (354 mV). The relatively lower OER performance of MET-Co/NF and MET-Cu/NF compared to MET-Fe/NF can be attributed to three main reasons. First, the Fe element is likely to offer unique advantages in the nickel foam OER catalyst system due to its inherent electronic properties and catalytic efficiency [47–50]. Second, MET-Fe is characterized by its inherent electrical conductivity with a value of 0.77×10^{-4} S cm^{-1} reported by Yaghi et al. [28], which is essential for catalysis. The projected density of states (PDOS) calculation results (Figure S6) of MET-Fe indicate that MET-Fe seems to be a semiconductor (conductivities in the range from 10^{-8} to 10^{2} S cm^{-1}) considering its conductivity value. However, MET-Co and MET-Cu exhibit insulating properties like most other MOFs. This difference in conductivity could adversely affect the electron mobility within the composite, especially when compared to the baseline performance of pure nickel foam. Thirdly, the interaction between the MOFs and the nickel foam substrate shows minimal surface growth for MET-Co and excessive coverage by MET-Cu, which could clog the nickel foam structure and affect its effectiveness. In contrast, MET-Fe forms a uniform and dense growth layer on the nickel foam surface, ensuring the full participation of MET-Fe as the active material without affecting the conductivity of the substrate and its interaction with the electrolyte. These results highlight not only the superior performance of MET-Fe/NF over commercial RuO_2 but also its impressive material uniqueness within the MET/NF series of composites.

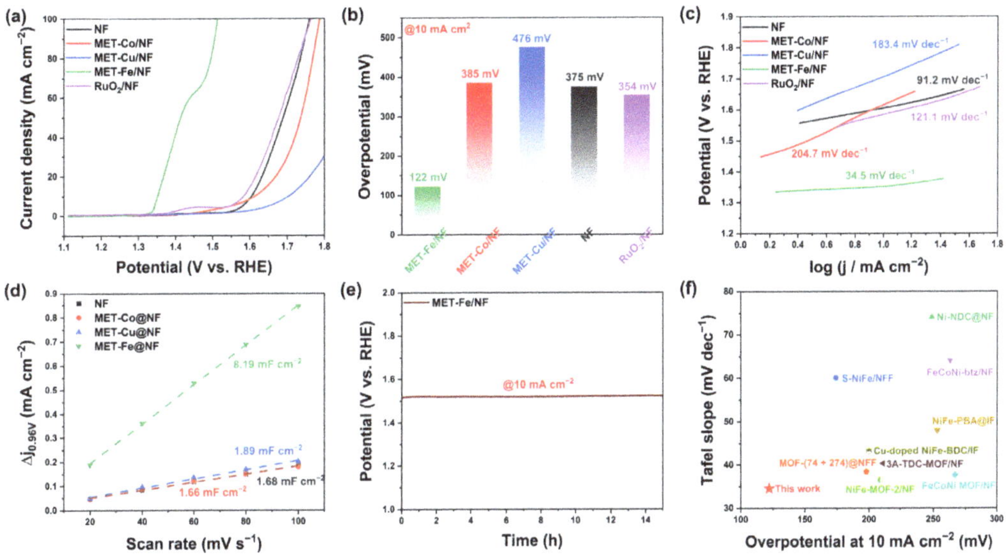

Figure 4. Electrocatalytic OER performance. (**a**) LSV curves. (**b**) Comparison of the overpotentials at 10 mA cm^{-2}. (**c**) Tafel plots. (**d**) Electrochemical double-layer capacitance. (**e**) Stability test at constant current density of 10 mA cm^{-2} without iR compensation. (**f**) Comparison of the OER activity of MET-Fe/NF and other reported catalysts: overpotential at 10 mA cm^{-2} and the corresponding Tafel slope.

Figure 4c reveals that the Tafel slope of MET-Fe/NF is 34.5 mV dec^{-1}, which is significantly lower compared to MET-Co/NF (204.7 mV dec^{-1}), MET-Cu/NF (183.4 mV dec^{-1}), bare nickel foam (NF) (91.2 mV dec^{-1}), and RuO_2/NF (121.1 mV dec^{-1}), highlighting the superior reaction kinetics of MET-Fe/NF. The electrochemical surface area (ECSA) serves as a key indicator of catalytic efficiency, closely linked to the double layer capacitance (C_{dl}). Cdl values were derived from the analysis of cyclic voltammetry (CV) curves across various scan rates within the non-Faradaic region, as shown in Figure S5. Illustrated in Figure 4d, the calculated C_{dl} values for NF, MET-Co/NF, MET-Cu/NF, and MET-Fe/NF are 1.68 mF cm^{-2}, 1.66 mF cm^{-2}, 1.89 mF cm^{-2}, and 8.19 mF cm^{-2}, respectively, with MET-Fe/NF exhibiting the highest C_{dl} value, indicating the highest number of active catalytic sites during the reaction.

The turnover frequency (TOF) is another important parameter to evaluate the electrocatalysts' activity [51]. The parameter was calculated with TOF = jA/4Fm, where j is current density, A is the surface area of the electrode, 4 indicates the number of electrons consumed for O_2 evolution from water, F is the Faraday constant, and m is the number of active species [52]. Figure S7 shows TOF curves at various overpotentials, indicating that the TOF values of all catalysts exhibited a monotonic increase with increasing overpotential. Among all other catalysts, the MET-Fe/NF exhibits the highest intrinsic catalytic activity at the same overpotential, suggesting the best OER performance.

Electrochemical impedance spectroscopy (EIS) facilitates the understanding of charge transfer at the reaction interface and provides Nyquist plots (Figure S8). The charge transfer resistance (R_{ct}) values derived from the Nyquist plots reveal that MET-Fe/NF has the smallest R_{ct} value of only 0.36 Ω, significantly lower than those of MET-Co/NF (3.97 Ω), MET-Cu/NF (3.86 Ω), NF (2.23 Ω), and RuO_2/NF (2.08 Ω). This suggests that MET-Fe/NF possesses superior charge transfer capabilities, which is conducive to improving the OER process. These results collectively demonstrate the exceptional OER catalytic activity of MET-Fe/NF compared to the other samples. Stability tests were also carried out on MET-Fe/NF at a current density of 10 mA cm^{-2}. The results indicate that MET-Fe/NF exhibits a mere performance degradation after 15 h of reaction, as reflected by the little change (+9 mV) in voltage values, showcasing remarkable stability (Figure 4e). A comparison of MET-Fe/NF with recently reported MOF/metal foam-based OER catalysts reveals superior performance in terms of both overpotential and Tafel slope [53–61] (Figure 4f). Taken together, these results establish MET-Fe/NF as an advanced OER catalyst that combines high catalytic activity with remarkable stability.

To assess the practical application performance of MET-Fe/NF further, it was combined with Pt/C/NF in a two-electrode system for a water-splitting experiment conducted in 1 M KOH electrolyte. The LSV curve of the MET-Fe/NF | | Pt/C/NF two-electrode electrolysis setup, shown in Figure 5a, demonstrates that this water-splitting apparatus needs just a low cell voltage of 1.463 V to reach an electrolysis current density of 10 mA cm^{-2}, highlighting its remarkable water-splitting efficiency and superiority to the commercial RuO_2-based two-electrode system (Figure S9). Insets in the figure illustrate the formation of gas bubbles on the electrode surfaces during the reaction, which correspond to oxygen and hydrogen gases generated by the anodic OER and the cathodic hydrogen evolution reaction (HER), respectively. The stability test results, depicted in Figure 5b, indicate that after 15 h of continuous operation at 10 mA cm^{-2}, splitting water into oxygen and hydrogen, the voltage required to drive the energy conversion device increased only by 20 mV, maintaining stable performance over a long period of time. These results highlight the efficient energy conversion efficiency and remarkable stability of MET-Fe/NF in water-splitting devices, demonstrating its broad application prospects in the large-scale production of hydrogen and oxygen gases.

To investigate the composition of the high catalytic activity sites in MET-Fe/NF for the OER, we characterized the sample after a prolonged series of tests using SEM, XRD, and XPS. The SEM images in Figure 6a reveal morphological changes in MET-Fe/NF after testing compared to its initial state, with several regions retaining their original morphology while others display a noticeable increase in uneven and rough surfaces. These rough surface areas are probably due to the structure reconstruction of the chemical composition that occurred during the OER process. The SEM images in Figure S10 indicate that the reconstructed new species consists of some large-sized nanoparticles in a blocky morphology on the surface of nickel foam. The XRD spectrum in Figure 6b provides clear evidence of material transformation and the emergence of new compounds within MET-Fe/NF throughout the testing. Peaks around 9.2° corresponding to the (1 1 1) plane of MET-Fe suggest that the post-testing sample still retains MET-Fe components. However, the appearance of numerous new peaks in the spectrum indicates the presence of a significant amount of metal hydroxides, including FeOOH (PDF# 22-0353), NiOOH (PDF# 27-0956), and nickel iron-layered double hydroxide (NiFe-LDH, PDF# 49-0188).

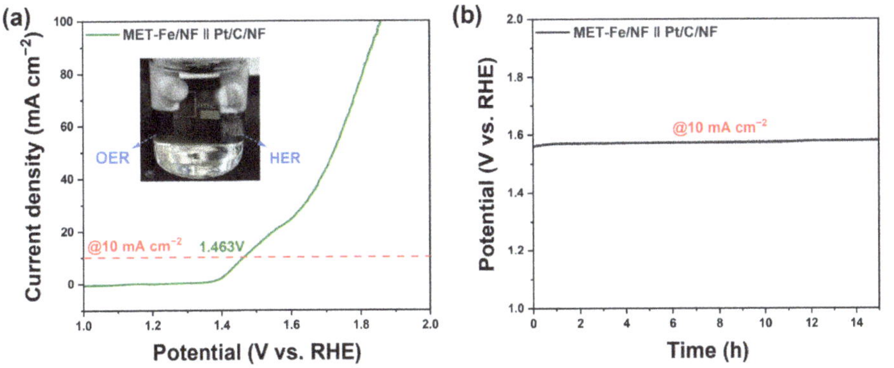

Figure 5. (**a**) LSV curve and photograph of overall water splitting over the MET-Fe/NF | | Pt/C/NF two-electrode setup. (**b**) Stability test of overall water splitting at constant current density of 10 mA cm^{-2} without iR compensation.

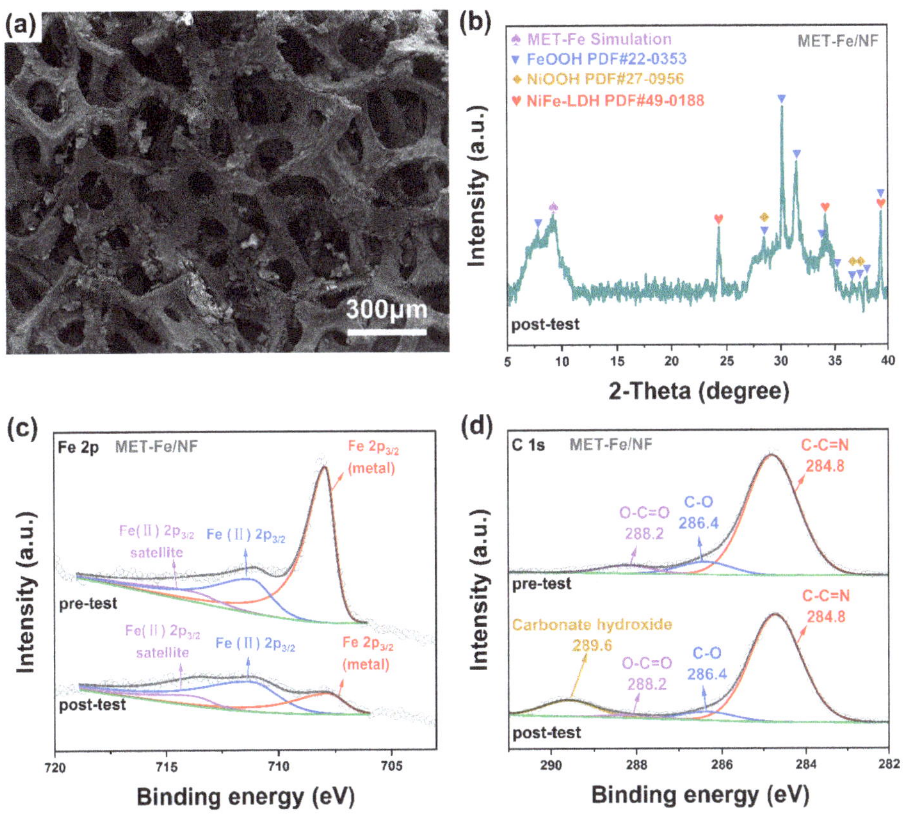

Figure 6. (**a**) SEM images of MET-Fe/NF after OER test. (**b**) XRD pattern of MET-Fe/NF after OER test. High-resolution XPS spectra of (**c**) Fe 2p and (**d**) C 1s of MET-Fe/NF before and after OER test.

XPS spectra further reveal the changes in chemical composition after the test. The full spectrum in Figure S10a shows the presence of Ni, Fe, O, N, C, and K elements in the post-test sample, with the newly added K element and a significant increase in the O element resulting from the adsorption of the electrolyte KOH after the test. The Fe 2p spectrum (Figure 6c) with a characteristic peak at 707.9 eV (metal state) shows a significant decrease in this component after the test, indicating that a large proportion of MET-Fe is transformed into other substances. The peak at 711.2 eV corresponding to higher oxidation states of Fe(II) [62,63] indicates the transformation of Fe into iron-based hydroxides during the structure reconstruction process [58,64]. In the C 1s spectrum (Figure 6d), the peak at 289.6 eV [65] corresponding to carbonates in layered double hydroxides confirms the formation of NiFe-LDH during the restructuring of MET-Fe/NF. An increase in the proportion of the C=O peak in the O 1s spectrum (Figure S11b) after the test and its shift towards lower binding energy also indicate the transformation of MET-Fe to NiFe-LDH. The FT-IR spectrum of MET-Fe/NF before and after the OER test also indicates the formation of metal oxyhydroxides and metal carbonate hydroxides.

In conclusion, the MET-Fe/NF underwent chemical restructuring after the test, with a significant proportion of the MET-Fe being converted to metal oxyhydroxides or metal carbonate hydroxides such as FeOOH and NiFe-LDH; the presence of NiOOH, with Ni originating from the nickel foam substrate, was also observed. These different metal hydroxides together form a unique hybrid OER catalyst, revealing the material composition responsible for the high OER catalytic activity of MET-Fe/NF.

The high OER activity of MET-Fe-NF originates from the following advantages: (1) iron/nickel (oxy)hydroxides reconstructing from MOFs have been confirmed to be highly efficient OER catalysts; (2) the synergy effect among multiple active components (FeOOH, NiOOH, and NiFe-LDH) enhances the OER performance of the hybrid catalysts; (3) nickel foam (NF) is a good conductive substrate that can considerably accelerate the electron and mass transfer of OER; (4) the high porosity and uniform growth of MET-Fe on the substrate make the reconstructed iron/nickel (oxy)hydroxides highly exposed for OER.

In fact, our work and the literature have gradually shown the poor stability of MOFs in extremely alkaline solutions and their restructuring in metal (oxy)hydroxides; however, this apparent issue might also provide novel synthetic approaches [66]. Lastly, it is important to emphasize that our synthesis technique is easy to use, safe, and economical, and produces catalysts with outstanding activity.

4. Conclusions

In this study, we developed the MOF/NF composites using a solvothermal method, directly employing it as a self-supporting electrode for the OER. MOF-coated nickel foam substrates offered superior conductivity and efficient exposure of the solid–liquid–gas interface due to its high porosity. The optimized catalyst MET-Fe/NF demonstrated the best OER performance, achieving a low overpotential of 122 mV at a current density of 10 mA cm^{-2} and exhibiting robust stability over 15 h. Used as the anode in a two-electrode water-splitting setup, MET-Fe/NF | | Pt/C/NF enabled continuous hydrogen and oxygen generation at an operating voltage of 1.463 V, highlighting its applicability in practical applications. During the OER process, MET-Fe/NF underwent a chemical transformation to a hybrid catalyst consisting of NiOOH, FeOOH, and NiFe-LDH, as verified by SEM, XRD, and XPS analyses. This transformation introduced a diverse range of active components recognized for their catalytic activity in OER. This study points to the possibility of easily fabricating self-supported MOF-based electrodes for electrocatalytic water splitting. In the future, more advanced in situ instruments should be employed in electrocatalysis to ensure the accurate design of efficient and stable electrocatalysts.

Supplementary Materials: The following supporting information can be downloaded at: https://www.mdpi.com/article/10.3390/nano14141168/s1, Figure S1: (a) Illustration of 1H-1,2,3-triazole molecule (**1**) deprotonates to form the triazolate anion. Every nitrogen atom (including N1 type and N2 type which are different from chemical environment) is able to coordinate several metal atoms in mode (**2**). (b) Illustration of MET-Fe with a pore (yellow sphere). Iron atoms are represented as brown spheres or polyhedron, nitrogen and carbon atoms are blue and grey spheres, respectively; Figure S2: The photograph of (a) NF, (b) MET-Fe/NF, (c) MET-Co/NF and (d) MET-Cu/NF; Figure S3: XRD patterns of (a) MET-Fe, (b) MET-Co/NF and MET-Cu/NF, (c) MET-Co and (d) MET-Cu; Figure S4: (a) SEM images of NF. (b) Elemental mapping images of NF. SEM images of (c) MET-Co/NF and (d) MET-Cu/NF; Figure S5: Cyclic voltammetry profiles at different scan rates (20–100 mV) in 1.0 M KOH for (a) MET-Fe/NF, (b) MET-Co/NF, (c) MET-Cu/NF and (d) NF; Figure S6: The projected density of states (PDOS) calculation of MET-Fe on the mainly involved elements and orbitals; Figure S7: The turnover frequency (TOF) profiles versus overpotential of MET-Fe/NF, MET-Co/NF and MET-Cu/NF; Figure S8: (a) Nyquist plots (The insert shows the equivalent circuit). (b) Charge transfer resistance based on the fitting data; Figure S9: LSV curve comparation of overall water splitting over the MET-Fe/NF||Pt/C/NF and RuO2/NF||Pt/C/NF two-electrode setups; Figure S10: SEM of MET-Fe/NF after OER test; Figure S11: XPS survey spectrum of (a) MET-Fe/NF after OER test. High resolution XPS spectra of (b) O 1s of MET-Fe/NF before and after OER test; Figure S12: FT-IR spectrum of MET-Fe/NF before and after OER test; Table S1: Comparison of OER performance with recently reported MOF-based superior electrocatalysts in 1M KOH; Table S2: Comparison of over-water splitting activity performance with recently reported MOF-based superior electrocatalysts in 1M KOH. References [67–72] are cited in the Supplementary Materials.

Author Contributions: Conceptualization, K.C.; Methodology, K.C.; Formal analysis, K.C.; Investigation, K.C. and Y.W.; Data curation, H.C.; Writing—original draft, W.C.; Writing—review & editing, W.C.; Visualization, W.C.; Supervision, X.H. and R.Z.; Project administration, W.C., X.H. and R.Z.; Funding acquisition, X.H. and R.Z. All authors have read and agreed to the published version of the manuscript.

Funding: This work was financially supported by the National Natural Science Foundation of China (52332007, 51825201 and 52227802).

Data Availability Statement: Data is contained within the article or Supplementary Materials.

Conflicts of Interest: The authors declare no competing financial interests.

References

1. Li, J.; Zhou, Z.; Xu, H.; Wang, C.; Hata, S.; Dai, Z.; Shiraishi, Y.; Du, Y. In situ nanopores enrichment of Mesh-like palladium nanoplates for bifunctional fuel cell reactions: A joint etching strategy. *J. Colloid Interface Sci.* **2022**, *611*, 523–532. [CrossRef]
2. Zhou, X.; Gao, J.; Hu, Y.; Jin, Z.; Hu, K.; Reddy, K.M.; Yuan, Q.; Lin, X.; Qiu, H.-J. Theoretically Revealed and Experimentally Demonstrated Synergistic Electronic Interaction of CoFe Dual-Metal Sites on N-doped Carbon for Boosting Both Oxygen Reduction and Evolution Reactions. *Nano Lett.* **2022**, *22*, 3392–3399. [CrossRef] [PubMed]
3. Zhai, P.; Xia, M.; Wu, Y.; Zhang, G.; Gao, J.; Zhang, B.; Cao, S.; Zhang, Y.; Li, Z.; Fan, Z.; et al. Engineering single-atomic ruthenium catalytic sites on defective nickel-iron layered double hydroxide for overall water splitting. *Nat. Commun.* **2021**, *12*, 4587. [CrossRef]
4. Li, T.-M.; Hu, B.-Q.; Han, J.-H.; Lu, W.; Yu, F.; Li, B. Highly Effective OER Electrocatalysts Generated from a Two-Dimensional Metal–Organic Framework Including a Sulfur-Containing Linker without Doping. *Inorg. Chem.* **2022**, *61*, 7051–7059. [CrossRef] [PubMed]
5. Li, D.; Wang, J.; Guo, S.; Xiao, Y.; Zeng, Q.; He, W.; Gan, L.; Zhang, Q.; Huang, S. Molecular-Scale Interface Engineering of Metal–Organic Frameworks toward Ion Transport Enables High-Performance Solid Lithium Metal Battery. *Adv. Funct. Mater.* **2020**, *30*, 2003945. [CrossRef]
6. Chan, C.K.; Peng, H.; Liu, G.; McIlwrath, K.; Zhang, X.F.; Huggins, R.A.; Cui, Y. High-performance lithium battery anodes using silicon nanowires. *Nat. Nanotechnol.* **2008**, *3*, 31–35. [CrossRef] [PubMed]
7. Xu, H.; Yuan, J.; He, G.; Chen, H. Current and future trends for spinel-type electrocatalysts in electrocatalytic oxygen evolution reaction. *Coord. Chem. Rev.* **2023**, *475*, 214869. [CrossRef]
8. Roger, I.; Shipman, M.A.; Symes, M.D. Earth-abundant catalysts for electrochemical and photoelectrochemical water splitting. *Nat. Rev. Chem.* **2017**, *1*, 0003. [CrossRef]
9. Reier, T.; Oezaslan, M.; Strasser, P. Electrocatalytic Oxygen Evolution Reaction (OER) on Ru, Ir, and Pt Catalysts: A Comparative Study of Nanoparticles and Bulk Materials. *ACS Catal.* **2012**, *2*, 1765–1772. [CrossRef]
10. Seitz, L.C.; Dickens, C.F.; Nishio, K.; Hikita, Y.; Montoya, J.; Doyle, A.; Kirk, C.; Vojvodic, A.; Hwang, H.Y.; Norskov, J.K.; et al. A highly active and stable IrOx/SrIrO3 catalyst for the oxygen evolution reaction. *Science* **2016**, *353*, 1011–1014. [CrossRef] [PubMed]
11. Lee, Y.; Suntivich, J.; May, K.J.; Perry, E.E.; Shao-Horn, Y. Synthesis and Activities of Rutile IrO_2 and RuO_2 Nanoparticles for Oxygen Evolution in Acid and Alkaline Solutions. *J. Phys. Chem. Lett.* **2012**, *3*, 399–404. [CrossRef] [PubMed]

12. Zhang, K.; Zhang, Z.; Shen, H.; Tang, Y.; Liang, Z.; Zou, R. Electronic modulation of Ni2P through anion and cation substitution toward highly efficient oxygen evolution. *Sci. China Mater.* **2022**, *65*, 1522–1530. [CrossRef]
13. Liang, Z.; Zhou, W.; Gao, S.; Zhao, R.; Zhang, H.; Tang, Y.; Cheng, J.; Qiu, T.; Zhu, B.; Qu, C.; et al. Fabrication of Hollow CoP/TiOx Heterostructures for Enhanced Oxygen Evolution Reaction. *Small* **2020**, *16*, 1905075. [CrossRef] [PubMed]
14. Zhang, L.; Guo, X.; Zhang, S.; Huang, S. Building up the "Genome" of bi-atom catalysts toward efficient HER/OER/ORR. *J. Mater. Chem. A* **2022**, *10*, 11600–11612. [CrossRef]
15. Xu, H.; Shang, H.; Wang, C.; Du, Y. Low-Dimensional Metallic Nanomaterials for Advanced Electrocatalysis. *Adv. Funct. Mater.* **2020**, *30*, 2006317. [CrossRef]
16. Zheng, S.; Li, Q.; Xue, H.; Pang, H.; Xu, Q. A highly alkaline-stable metal oxide@metal–organic framework composite for high-performance electrochemical energy storage. *Natl. Sci. Rev.* **2019**, *7*, 305–314. [CrossRef] [PubMed]
17. Liu, J.; Wöll, C. Surface-supported metal–organic framework thin films: Fabrication methods, applications, and challenges. *Chem. Soc. Rev.* **2017**, *46*, 5730–5770. [CrossRef]
18. Furukawa, H.; Cordova, K.E.; O'Keeffe, M.; Yaghi, O.M. The Chemistry and Applications of Metal-Organic Frameworks. *Science* **2013**, *341*, 1230444. [CrossRef]
19. Qian, Q.; Li, Y.; Liu, Y.; Yu, L.; Zhang, G. Ambient Fast Synthesis and Active Sites Deciphering of Hierarchical Foam-Like Trimetal–Organic Framework Nanostructures as a Platform for Highly Efficient Oxygen Evolution Electrocatalysis. *Adv. Mater.* **2019**, *31*, 1901139. [CrossRef] [PubMed]
20. Dong, H.; Zhang, X.; Yan, X.-C.; Wang, Y.-X.; Sun, X.; Zhang, G.; Feng, Y.; Zhang, F.-M. Mixed-Metal-Cluster Strategy for Boosting Electrocatalytic Oxygen Evolution Reaction of Robust Metal–Organic Frameworks. *ACS Appl. Mater. Interfaces* **2019**, *11*, 45080–45086. [CrossRef] [PubMed]
21. Zhang, B.; Zheng, Y.; Ma, T.; Yang, C.; Peng, Y.; Zhou, Z.; Zhou, M.; Li, S.; Wang, Y.; Cheng, C. Designing MOF Nanoarchitectures for Electrochemical Water Splitting. *Adv. Mater.* **2021**, *33*, 2006042. [CrossRef]
22. Poudel, M.B.; Vijayapradeep, S.; Sekar, K.; Kim, J.S.; Yoo, D.J. Pyridinic-N exclusively enriched CNT-encapsulated NiFe interfacial alloy nanoparticles on knitted carbon fiber cloth as bifunctional oxygen catalysts for biaxially flexible zinc–air batteries. *J. Mater. Chem. A* **2024**, *12*, 10185–10195. [CrossRef]
23. Xu, X.; Sun, H.; Jiang, S.P.; Shao, Z. Modulating metal–organic frameworks for catalyzing acidic oxygen evolution for proton exchange membrane water electrolysis. *SusMat* **2021**, *1*, 460–481. [CrossRef]
24. Pan, S.; Ma, Z.; Yang, W.; Dongyang, B.; Yang, H.; Lai, S.; Dong, F.; Yang, X.; Lin, Z. Magnesium incorporation activates perovskite cobaltites toward efficient and stable electrocatalytic oxygen evolution. *Mater. Rep. Energy* **2023**, *3*, 100212. [CrossRef]
25. Wang, H.-F.; Chen, L.; Pang, H.; Kaskel, S.; Xu, Q. MOF-derived electrocatalysts for oxygen reduction, oxygen evolution and hydrogen evolution reactions. *Chem. Soc. Rev.* **2020**, *49*, 1414–1448. [CrossRef] [PubMed]
26. Lu, X.F.; Xia, B.Y.; Zang, S.-Q.; Lou, X.W. Metal–Organic Frameworks Based Electrocatalysts for the Oxygen Reduction Reaction. *Angew. Chem. Int. Ed.* **2020**, *59*, 4634–4650. [CrossRef] [PubMed]
27. Zou, Z.; Wang, T.; Zhao, X.; Jiang, W.-J.; Pan, H.; Gao, D.; Xu, C. Expediting in-Situ Electrochemical Activation of Two-Dimensional Metal–Organic Frameworks for Enhanced OER Intrinsic Activity by Iron Incorporation. *ACS Catal.* **2019**, *9*, 7356–7364. [CrossRef]
28. Gándara, F.; Uribe-Romo, F.J.; Britt, D.K.; Furukawa, H.; Lei, L.; Cheng, R.; Duan, X.; O'Keeffe, M.; Yaghi, O.M. Porous, Conductive Metal-Triazolates and Their Structural Elucidation by the Charge-Flipping Method. *Chem.—Eur. J.* **2012**, *18*, 10595–10601. [CrossRef] [PubMed]
29. Cao, C.; Ma, D.-D.; Xu, Q.; Wu, X.-T.; Zhu, Q.-L. Semisacrificial Template Growth of Self-Supporting MOF Nanocomposite Electrode for Efficient Electrocatalytic Water Oxidation. *Adv. Funct. Mater.* **2019**, *29*, 1807418. [CrossRef]
30. Song, X.Z.; Zhang, N.; Wang, X.F.; Tan, Z. Recent advances of metal-organic frameworks and their composites toward oxygen evolution electrocatalysis. *Mater. Today Energy* **2021**, *19*, 100597. [CrossRef]
31. Rui, K.; Zhao, G.; Chen, Y.; Lin, Y.; Zhou, Q.; Chen, J.; Zhu, J.; Sun, W.; Huang, W.; Dou, S.X. Hybrid 2D Dual-Metal–Organic Frameworks for Enhanced Water Oxidation Catalysis. *Adv. Funct. Mater.* **2018**, *28*, 1801554. [CrossRef]
32. Cheng, W.; Xi, S.; Wu, Z.-P.; Luan, D.; Lou, X.W. In situ activation of Br-confined Ni-based metal-organic framework hollow prisms toward efficient electrochemical oxygen evolution. *Sci. Adv.* **2021**, *7*, eabk0919. [CrossRef] [PubMed]
33. Tian, J.; Jiang, F.; Yuan, D.; Zhang, L.; Chen, Q.; Hong, M. Electric-Field Assisted In Situ Hydrolysis of Bulk Metal–Organic Frameworks (MOFs) into Ultrathin Metal Oxyhydroxide Nanosheets for Efficient Oxygen Evolution. *Angew. Chem. Int. Ed.* **2020**, *59*, 13101–13108. [CrossRef] [PubMed]
34. Zhe, T.; Li, F.; Ma, K.; Liu, M.; Li, R.; Li, M.; Wang, C.; Luo, Q.; Lü, X.; Wang, L. Accelerated Oxygen Evolution Kinetics by Engineering Heterojunction Coupling of Amorphous NiFe Hydr(oxy)oxide Nanosheet Arrays on Self-Supporting Ni-MOFs. *Small* **2023**, *19*, 2303303. [CrossRef] [PubMed]
35. Zheng, W.; Liu, M.; Lee, L.Y.S. Electrochemical Instability of Metal–Organic Frameworks: In Situ Spectroelectrochemical Investigation of the Real Active Sites. *ACS Catal.* **2020**, *10*, 81–92. [CrossRef]
36. Zhao, S.; Tan, C.; He, C.-T.; An, P.; Xie, F.; Jiang, S.; Zhu, Y.; Wu, K.-H.; Zhang, B.; Li, H.; et al. Structural transformation of highly active metal–organic framework electrocatalysts during the oxygen evolution reaction. *Nat. Energy* **2020**, *5*, 881–890. [CrossRef]
37. Zhong, H.; Zhang, Q.; Yu, J.; Zhang, X.; Wu, C.; Ma, Y.; An, H.; Wang, H.; Zhang, J.; Wang, X.; et al. Fundamental Understanding of Structural Reconstruction Behaviors in Oxygen Evolution Reaction Electrocatalysts. *Adv. Energy Mater.* **2023**, *13*, 2301391. [CrossRef]

38. Huang, J.; Li, Y.; Zhang, Y.; Rao, G.; Wu, C.; Hu, Y.; Wang, X.; Lu, R.; Li, Y.; Xiong, J. Identification of Key Reversible Intermediates in Self-Reconstructed Nickel-Based Hybrid Electrocatalysts for Oxygen Evolution. *Angew. Chem. Int. Ed.* **2019**, *58*, 17458–17464. [CrossRef]
39. Li, X.; Fang, Y.; Lin, X.; Tian, M.; An, X.; Fu, Y.; Li, R.; Jin, J.; Ma, J. MOF derived Co3O4 nanoparticles embedded in N-doped mesoporous carbon layer/MWCNT hybrids: Extraordinary bi-functional electrocatalysts for OER and ORR. *J. Mater. Chem. A* **2015**, *3*, 17392–17402. [CrossRef]
40. Chen, P.; Zhou, T.; Xing, L.; Xu, K.; Tong, Y.; Xie, H.; Zhang, L.; Yan, W.; Chu, W.; Wu, C.; et al. Atomically Dispersed Iron–Nitrogen Species as Electrocatalysts for Bifunctional Oxygen Evolution and Reduction Reactions. *Angew. Chem. Int. Ed.* **2017**, *56*, 610–614. [CrossRef]
41. Zhang, N.; Hu, Y.; An, L.; Li, Q.; Yin, J.; Li, J.; Yang, R.; Lu, M.; Zhang, S.; Xi, P.; et al. Surface Activation and Ni-S Stabilization in NiO/NiS2 for Efficient Oxygen Evolution Reaction. *Angew. Chem. Int. Ed.* **2022**, *61*, e202207217. [CrossRef] [PubMed]
42. Guan, D.; Ryu, G.; Hu, Z.; Zhou, J.; Dong, C.-L.; Huang, Y.-C.; Zhang, K.; Zhong, Y.; Komarek, A.C.; Zhu, M.; et al. Utilizing ion leaching effects for achieving high oxygen-evolving performance on hybrid nanocomposite with self-optimized behaviors. *Nat. Commun.* **2020**, *11*, 3376. [CrossRef] [PubMed]
43. Chu, S.; Sun, H.; Chen, G.; Chen, Y.; Zhou, W.; Shao, Z. Rationally designed Water-Insertable Layered Oxides with Synergistic Effect of Transition-Metal Elements for High-Performance Oxygen Evolution Reaction. *ACS Appl. Mater. Interfaces* **2019**, *11*, 25227–25235. [CrossRef] [PubMed]
44. Wang, X.; Wang, X.; Zhao, L.; Zhang, H.; Liu, M.; Zhang, C.; Liu, S. Self-reconstruction of cationic activated Ni-MOFs enhanced the intrinsic activity of electrocatalytic water oxidation. *Inorg. Chem. Front.* **2022**, *9*, 179–185. [CrossRef]
45. Chang, J.; Chen, L.; Zang, S.; Wang, Y.; Wu, D.; Xu, F.; Jiang, K.; Gao, Z. The effect of Fe(III) cations in electrolyte on oxygen evolution catalytic activity of Ni(OH)2 electrode. *J. Colloid Interface Sci.* **2020**, *569*, 50–56. [CrossRef] [PubMed]
46. Xu, Z.; Yeh, C.-L.; Jiang, Y.; Yun, X.; Li, C.-T.; Ho, K.-C.; Lin, J.T.; Lin, R.Y.-Y. Orientation-Adjustable Metal–Organic Framework Nanorods for Efficient Oxygen Evolution Reaction. *ACS Appl. Mater. Interfaces* **2021**, *13*, 28242–28251. [CrossRef]
47. Trotochaud, L.; Young, S.L.; Ranney, J.K.; Boettcher, S.W. Nickel–Iron Oxyhydroxide Oxygen-Evolution Electrocatalysts: The Role of Intentional and Incidental Iron Incorporation. *J. Am. Chem. Soc.* **2014**, *136*, 6744–6753. [CrossRef]
48. Corrigan, D.A. The Catalysis of the Oxygen Evolution Reaction by Iron Impurities in Thin Film Nickel Oxide Electrodes. *J. Electrochem. Soc.* **1987**, *134*, 377. [CrossRef]
49. Ou, Y.; Twight, L.P.; Samanta, B.; Liu, L.; Biswas, S.; Fehrs, J.L.; Sagui, N.A.; Villalobos, J.; Morales-Santelices, J.; Antipin, D.; et al. Cooperative Fe sites on transition metal (oxy)hydroxides drive high oxygen evolution activity in base. *Nat. Commun.* **2023**, *14*, 7688. [CrossRef]
50. Hu, J.; Guo, T.; Zhong, X.; Li, J.; Mei, Y.; Zhang, C.; Feng, Y.; Sun, M.; Meng, L.; Wang, Z.; et al. In Situ Reconstruction of High-Entropy Heterostructure Catalysts for Stable Oxygen Evolution Electrocatalysis under Industrial Conditions. *Adv. Mater.* **2024**, *36*, 2310918. [CrossRef]
51. Zhou, J.; Dou, Y.; Wu, X.-Q.; Zhou, A.; Shu, L.; Li, J.-R. Alkali-Etched Ni(II)-Based Metal–Organic Framework Nanosheet Arrays for Electrocatalytic Overall Water Splitting. *Small* **2020**, *16*, 1906564. [CrossRef] [PubMed]
52. Wang, Z.; Xu, J.; Yang, J.; Xue, Y.; Dai, L. Ultraviolet/ozone treatment for boosting OER activity of MOF nanoneedle arrays. *Chem. Eng. J.* **2022**, *427*, 131498. [CrossRef]
53. Jiang, Y.; Chen, T.-Y.; Chen, J.-L.; Liu, Y.; Yuan, X.; Yan, J.; Sun, Q.; Xu, Z.; Zhang, D.; Wang, X.; et al. Heterostructured Bimetallic MOF-on-MOF Architectures for Efficient Oxygen Evolution Reaction. *Adv. Mater.* **2023**, *36*, 2306910. [CrossRef] [PubMed]
54. Wu, L.; Feng, J.; Zou, Z.; Song, K.; Zeng, C. Formation of feathery-shaped dual-function S-doped FeNi-MOFs to achieve advanced electrocatalytic activity for OER and HER. *J. Electroanal. Chem.* **2023**, *935*, 117365. [CrossRef]
55. Wei, L.; Meng, D.; Mao, J.; Jiang, Q.; Huang, H.; Tang, J. Assembly of NiFe-PBA nanoparticles on nanoflower-like NiFe-PBA@IF as enhanced oxygen evolution electrocatalyst at room temperature. *Mol. Catal.* **2023**, *544*, 113126. [CrossRef]
56. Hu, F.; Yu, D.; Zeng, W.-J.; Lin, Z.-Y.; Han, S.; Sun, Y.; Wang, H.; Ren, J.; Hung, S.-F.; Li, L.; et al. Active Site Tailoring of Metal-Organic Frameworks for Highly Efficient Oxygen Evolution. *Adv. Energy Mater.* **2023**, *13*, 2301224. [CrossRef]
57. Li, W.; Jia, S.; Liu, X.; Li, Y.; Chen, T.; Yang, F.; Zhang, X. Exploration and Application of the Self-Optimization Phenomenon of a Trimetal-Based MOF Electrocatalyst in the Oxygen Evolution Reaction. *Energy Fuels* **2023**, *37*, 8563–8572. [CrossRef]
58. Wang, Q.; Ma, X.; Ma, P.; Bi, R.; Song, S. Defect-Rich, Rose-Shaped Fe2Ni1-Metal–Organic Framework Nanoarrays for Efficient Oxygen Evolution Reaction. *ACS Appl. Nano Mater.* **2023**, *6*, 9339–9350. [CrossRef]
59. Li, S.; Wang, Z.; Wang, T.; Yang, Y.; Xiao, Y.; Tian, Y.; Zhu, H.; Jing, X.; Zhu, G. Preparation of Trimetallic-Organic Framework Film Electrodes via Secondary Growth for Efficient Oxygen Evolution Reaction. *Chem.—Eur. J.* **2023**, *29*, e202301129. [CrossRef] [PubMed]
60. Wu, F.; Li, Q.; Guo, H.; Wang, S.; Hao, G.; Hu, Y.; Zhang, G.; Jiang, W. Modulating crystal and electronic structure of NiFe-MOFs by inorganic acid for highly efficient electrochemical water oxidation. *Dalton Trans.* **2023**, *52*, 2027–2035. [CrossRef] [PubMed]
61. Lee, M.K.; Choi, S.; Park, H.; Lee, T.H.; Lee, S.A.; Yang, J.W.; Ji, S.G.; Cheon, W.S.; Ahn, S.H.; Kim, S.Y.; et al. 2D Ni-Naphthalene-2,6-Dicarboxylic Acid Metal-Organic Framework as Electrocatalysts for Efficient Overall Water Splitting. *Energy Technol.* **2023**, *11*, 2201203. [CrossRef]

62. Biesinger, M.C.; Payne, B.P.; Grosvenor, A.P.; Lau, L.W.M.; Gerson, A.R.; Smart, R.S.C. Resolving surface chemical states in XPS analysis of first row transition metals, oxides and hydroxides: Cr, Mn, Fe, Co and Ni. *Appl. Surf. Sci.* **2011**, *257*, 2717–2730. [CrossRef]
63. Uhlig, I.; Szargan, R.; Nesbitt, H.W.; Laajalehto, K. Surface states and reactivity of pyrite and marcasite. *Appl. Surf. Sci.* **2001**, *179*, 222–229. [CrossRef]
64. Wang, M.; Chen, Y.; Yu, Z.; Hou, Y.; Jiang, R.; Li, S.; Chen, J.; Tang, W.; Pang, H.; Xie, W. Unraveling the π-interaction of NiFe-based metal–organic frameworks with enhanced oxygen evolution: Optimizing electronic structure and facilitating electron transfer modulation. *J. Colloid Interface Sci.* **2023**, *640*, 1–14. [CrossRef] [PubMed]
65. Shchukarev, A.V.; Korolkov, D.V. XPS Study of group IA carbonates. *Cent. Eur. J. Chem.* **2004**, *2*, 347–362. [CrossRef]
66. Tang, P.; Di Vizio, B.; Yang, J.; Patil, B.; Cattelan, M.; Agnoli, S. Fe,Ni-Based Metal–Organic Frameworks Embedded in Nanoporous Nitrogen-Doped Graphene as a Highly Efficient Electrocatalyst for the Oxygen Evolution Reaction. *Nanomaterials* **2024**, *14*, 751. [CrossRef] [PubMed]
67. Liu, J.; Yu, Z.; Huang, J.; Yao, S.; Jiang, R.; Hou, Y.; Tang, W.; Sun, P.; Huang, H.; Wang, M. Redox-active ligands enhance oxygen evolution reaction activity: Regulating the spin state of ferric ions and accelerating electron transfer. *J. Colloid Interface Sci.* **2023**, *650*, 1182–1192. [CrossRef] [PubMed]
68. Zhao, S.; Deng, L.; Xiong, Y.; Hu, F.; Yin, L.; Yu, D.; Li, L.; Peng, S. Engineering metal-organic framework nanosheets with electronically modulated in-plane heterojunctions for robust high-current-density water splitting. *Sci. China Mater.* **2023**, *66*, 1373–1382. [CrossRef]
69. Yin, Z.; Liang, J.; Zhang, Z.; Luo, H.; Zhou, J. Construction of superhydrophilic metal-organic frameworks with hierarchical microstructure for efficient overall water splitting. *J. Colloid Interface Sci.* **2022**, *623*, 405–416. [CrossRef] [PubMed]
70. Van Phuc, T.; Jana, J.; Ravi, N.; Kang, S.G.; Chung, J.S.; Choi, W.M.; Hur, S.H. Highly active Ni/Co-metal organic framework bifunctional electrocatalyst for water splitting reaction. *Int. J. Hydrog. Energy* **2022**, *47*, 22787–22795. [CrossRef]
71. Liu, H.; Zhang, T.; Cui, D.; Zheng, Y.; Cheng, Y.; Wang, G.; Chen, L. Defective ferrocene-based metal–organic frameworks for efficient solar-powered water oxidation via the ligand competition and etching effect. *J. Colloid Interface Sci.* **2024**, *657*, 664–671. [CrossRef] [PubMed]
72. Zhang, S.; Huang, Z.; Isimjan, T.T.; Cai, D.; Yang, X. Accurately substituting Fe for Ni2 atom in Ni-MOF with defect-rich for efficient oxygen evolution reaction: Electronic reconfiguration and mechanistic study. *Appl. Catal. B Environ.* **2024**, *343*, 123448. [CrossRef]

MDPI AG
Grosspeteranlage 5
4052 Basel
Switzerland
Tel.: +41 61 683 77 34

Nanomaterials Editorial Office
E-mail: nanomaterials@mdpi.com
www.mdpi.com/journal/nanomaterials

www.ingramcontent.com/pod-product-compliance
Lightning Source LLC
Chambersburg PA
CBHW062054220926
51850CB00010B/2761
* 9 7 8 3 7 2 5 8 6 3 9 2 1 *